OUR ONLY HOME

OUR ONLY HOME

A Climate Appeal

His Holiness the Dalai Lama
with Franz Alt

BLOOMSBURY SIGMA
LONDON · OXFORD · NEW YORK · NEW DELHI · SYDNEY

BLOOMSBURY SIGMA
Bloomsbury Publishing Plc
50 Bedford Square, London, WC1B 3DP, UK

BLOOMSBURY, BLOOMSBURY SIGMA and the Diana logo are trademarks
of Bloomsbury Publishing Plc

First published in the United Kingdom by Bloomsbury Publishing 2020
First published in Germany by Bonavento in 2019
© Benevento by Benevento Publishing

The Dalai Lama and Franz Alt have asserted their rights under the Copyright,
Designs and Patents Act, 1988, to be identified as Author of this work

A catalogue record for this book is available from the British Library

Library of Congress Cataloguing-in-Publication data has been applied for

ISBN: HB: 978-1-4729-8392-3; eBook: 978-1-4729-8075-5

2 4 6 8 10 9 7 5 3 1

Typeset in Fournier MT Std by Deanta Global Publishing Services, Chennai, India
Printed and bound in Great Britain by CPI Group (UK) Ltd, Croydon CR0 4YY

To find out more about our authors and books visit www.bloomsbury.com
and sign up for our newsletters

CONTENTS

Introduction
by Franz Alt

1. Life is holy

'I am also an ardent supporter of environmental protection. We humans are the only species with the power to destroy the Earth as we know it. Yet, if we have the capacity to destroy the Earth, so, too, do we have the capacity to protect it.

'It is encouraging to see how you have opened the eyes of the world to the urgency to protect our planet, our only home. At the same time, you have inspired so many young brothers and sisters to join this movement.'

This is what the Dalai Lama wrote to the sixteen-year-old Swedish climate activist Greta Thunberg on 31 May 2019.

In the following months, Thunberg was received by the Pope and by former US President Barack Obama. She spoke before the United Nations (UN), before the French Parliament and at two world climate summits, as well as at the World Economic Forum in Davos. She was awarded the Alternative Nobel Prize, was invited by the US Senate to speak and was honoured with Amnesty International's Ambassador of Conscience award. But what had really changed? Back then, when young people took to the streets every Friday, where were the adults?

Barack Obama said to the shy, calm and serious young lady: 'You and me, we're a team'. Her simple answer: 'Yes'. Her motto seems to be: be humble. Before the UN summit that September, however, she flung her anger at politicians across the world with tears and in a trembling voice: 'You have stolen my childhood. You are failing us. People are suffering, people are dying, entire ecosystems are collapsing.'

With clenched fist she went on: 'We are in the beginning of a mass extinction, and all you can talk about is money and fairy tales of eternal economic growth. How dare you continue to look away and come here and say you're doing enough when the politics and solutions needed are still nowhere in sight?'

Greta's curse! For a moment, the politicians had become students to whom the riot act was being read, and the student had become the teacher.

Politicians are old pros, however. In Germany, their contribution was to suggest Saturday demonstrations rather than Fridays, a more traditional approach to protest that does not disrupt the working week. Such a complacent reaction showed how powerful courageously spoken truth can be. Young climate protesters see through politicians who sell their responsibility to benefit the profits of large companies.

Before the global pandemic, outrage expressed by adults at the destruction of our planet was far from being loud enough. Perhaps it will remain too quiet when the pandemic ends. Global warming is a worldwide catastrophe unprecedented in human history, because we are pursuing growth for the sake of growth. The result is that we are growing poorer. While economic growth increases, there is no matching increase in our overall prosperity. We have forgotten to ask: growth, for what and for whom? We have become blind to the ecological consequences of the drive for growth.

Greta Thunberg and her followers aim to wake us up. Maybe just in time.

Once the pandemic passes, a tsunami is still rolling towards us. But many of us continue to close our eyes, plug our ears and cover our mouths in the face of the danger, like the three famous Japanese monkeys.

After Greta's speech at the United Nations, the magazine *Der Spiegel* wondered: 'Could this be the only sensible person in a crazy world?' It suggested that her words might one day be considered 'a key speech of the early twenty-first century'.

2. Climate emergency

The Covid-19 pandemic only distracted people from a serious threat that was already looming on the horizon. In the first half of 2019 alone, millions of people in the developing world lost their homes and belongings due to the effects of global warming, the poorest in particular. How can a religious leader and spiritual teacher help in such a situation?

Over the past thirty-eight years, I have been lucky to meet the Dalai Lama 40 times and record 15 television interviews with him about peace, human rights and environmental and climate protection. This book is a continuation of that partnership. In it, the Dalai Lama appeals to the world to support the young climate activists to take a more active role in protecting this planet and to politicians to urgently tackle the global warming caused by climate change.

Besides Greta, other wise and courageous women, such as the Kenyan scientist and Nobel Peace Prize laureate Wangari Maathei or Indian agronomist and Alternative Nobel Prize winner Voandana Shiva, motivate us to implement ecological change and develop a sustainable market economy based on the philosophy 'eco-social instead of radical free-market'.

In my conversation with the Dalai Lama before the pandemic, it was against the background of such ecological and spiritual concerns that he emphasised that we must re-examine what we have inherited, what we are responsible for ... and what we will pass on to coming generations.

3. The world's most likeable person

Seldom in 50 years as a journalist have I spoken with such an empathetic, likeable and humorous person as the Dalai Lama. No one has laughed more than he did. It is no coincidence that surveys show he is widely considered the happiest person worldwide. To this religious leader, interreligious differences have become increasingly important in recent years. And what he says today distinguishes him from other religious leaders: 'Ethics is more important than religion. We are not born a member of a particular religion. But ethics is innate in all of us.' In his lectures worldwide, he speaks more and more frequently about 'secular ethics beyond all religions'. Albert Schweitzer called the same concern 'reverence for all life'.

The Dalai Lama's code of secular or ecological ethics breaks down national, religious and cultural boundaries and outlines values that are innate in and binding upon everyone. These are not external, material values, but inner values such as mindfulness, compassion for all creatures and mental training, as well as the pursuit of happiness. 'There is no justice without compassion and mercy', says the Dalai Lama. 'If we want to be happy ourselves, we should practice compassion, and if we want others to be happy, we should also practice compassion. We all prefer to see smiling rather than glum faces.'

One of the Dalai Lama's central beliefs is that, in our desire to pursue happiness and avoid suffering, we all equally wish to live a meaningful life. In the past, this has brought about humankind's greatest achievements. We should now be thinking and acting on the basis of deeper human values based on a sense of the oneness of humanity, with the aim of creating a more compassionate society.

Our conversations rest on the belief that all life is holy. Therefore, human dignity is the highest individual value and public welfare the highest collective value.

4. Humankind's survival is at stake

In a time of pandemic, global warming, extinction of species and increasing water emergency, the values of international cooperation and 'universal responsibility' for which the Dalai Lama calls ever more urgently are particularly important. In this book, as never before, he urges politicians to act decisively after more than 20 inconclusive international climate conferences, since the stakes are nothing less than the survival of our planet and the sanctity of life. Already, there are regions on Earth that are scarcely reminiscent of our good old world.

It has been the vision of the Dalai Lama to turn his home country, Tibet, into the world's largest nature reserve, in accordance with the ancient Tibetan

Buddhist tradition: 'Tibet must and can become a demilitarised sanctuary of peace and nature.'

Technology alone will not save us. Only if we combine technology with ethical responsibility can we – perhaps! – still prevent the worst effects of global warming. Yesterday, many parts of our beautiful blue planet were still natural paradises. Today, Earth is already withered in many places. By tomorrow, many regions will be uninhabitable if we just carry on as before. But there are always alternatives. All the problems created by people can also be solved by people.

2

The Third World War Against Nature
by Franz Alt

1. Humankind is losing control

What does our home planet look like in the year 2020?

The world is in the grip of a pandemic, which naturally commands everyone's attention. But once Covid-19 has passed – as it will – problems with which we're already familiar will remain. Rainforests everywhere will be on fire, deserts will be spreading on all continents, icebergs will be melting, global warming will mean millions of climate refugees. Are we beyond hope?

In 2019, after four years of dry weather, Australia experienced the worst drought in its history. Some villages in New South Wales now have to be supplied with drinking water from outside. Seawater desalination plants are already supplying a quarter of Sydney, while India is groaning under temperatures above 50 °C (122 °F). In Europe, thousands of elderly people died of the second heat wave after the very hot summer of 2018. In Brazil, in August 2019, twice as much forest was burning as the year before, and the deforestation rate was 222 percent higher than in August 2018. Brazilian President Jair Messias Bolsonaro – widely condemned for his response to the pandemic – calls Catholic bishops and priests who oppose this arson 'a rotten part of the Catholic Church'.

Autumn 2018 brought the most expensive disaster of that year worldwide and the worst fire in US history. It was both sad and ironic that the Californian city that burned down was called Paradise: eighty-five people lost their lives, more than 18,000 houses and buildings on 62,000 hectares were destroyed. What a symbol: 'Paradise' became a flaming hell, then a ghost town. Three hundred thousand people had to be evacuated. The damage amounted to about $14 billion.

Earth experiences devastating fires, heat-related deaths, drowning climate refugees, dying seas, rising sea levels, climate conflicts, polluted air, prohibitions on driving, economic collapse, water catastrophes and, of course, the pandemic. Global warming is far worse than we are willing to admit. The idea of a slowly changing climate is yesterday's fairy tale. The change is lightning fast. No place on Earth will remain untouched and no life will be unchanged. We are experiencing the greatest mass extinction of the past 65 million years.

When I was born in 1938, the climate system still seemed intact; today, it is completely out of control. This destruction was caused by we humans, who are behaving like pyromaniacs. In the span of a lifetime, we have driven our planet to the edge of the abyss. The one million refugees who arrived in Germany in 2015 were displaced because, in their homelands, climate change and drought were among the catalysts of civil war.

The environment is on its last legs, but in Germany car manufacturers report record sales of large SUVs. Germans love both: nature (the forest in particular) and the car. And it is not just Germans who have this ambivalent attitude: half the world is caught in the car trap.

Since 1945, 120 million people worldwide have been killed by cars. That is twice the deaths in the Second World War. And there's really no alternative? I haven't had a car for 10 years now. I make 98 percent of my journeys by public transport. I'm 100 times safer than in a car, I find time on the train to write my books and articles – and I also help the environment. Even the most modern car consumes resources that are lost forever; it costs an awful lot of money, yet it stands still for around 90 percent of the time, rusting away. Our automobiles are largely stationary, the very opposite of smart mobility. Without ecologically conscious cars, there will be no turnaround in energy policy (I can easily imagine that we'll share electric cars in the future).

'Humankind is losing control of the state of the Earth', warns Stefan Rahmstorf, co-chair of the Research Department of Earth System Analysis at the Potsdam Institute for Climate Impact Research, and advisor to the German Chancellor. Climate-research scientists have been mistaken on only one point during the past few decades: the climatic disaster is approaching much faster than they predicted.

Glaciologists admit that the ice today is melting three times as fast as they feared just ten years ago. In this century, sea levels are rising not by a few centimetres but by some metres. That means there will be even more sea. Which in turn means that not only half of Bangladesh will be uninhabitable, but also New York and Shanghai, Hamburg and Bremen, Mumbai and Calcutta, Alexandria and Rio. One in four of all Africans lives on the coast and will lose the ground beneath his or her feet if we do not stop global warming. The latest report of the Intergovernmental Panel on Climate Change predicts that flood damage will worsen by 100 to 1,000 times by the end of this century. The rising sea level is the sword of Damocles of global warming.

Worldwide, coral reefs are dying faster than predicted. The rate at which species are becoming extinct is breathtaking. Every day we are losing 150 animal and plant species, says renowned US biology Professor E. O. Wilson. We are the first generation to mess with God's creation. We are playing evolution backwards.

Glaciologists expect a rise in sea levels of up to 70 metres (230 ft), should the entire Greenland ice melt.

The goal of the 2015 Paris Climate Summit is to keep warming within 1.5 °C (34.7 °F) of pre-industrial levels will never be met if we continue at the current rate. Right now we are on track for up to 5 °C (41 °F), which means 46 to 48 °F degrees on land, leading to an African climate in southern Europe. There are still politicians and journalists who want to deny the facts provided by climate research or dismiss them as 'alarmist'. But changing language will not stop global warming. It is not a matter of faith, but of simple physics and science.

The fact is that, since humans are part of nature, we are waging a Third World War against ourselves. Professor Rahmstorf again: 'While climate researchers have been projecting global warming quite correctly for half a century, they were wrong concerning the pace and magnitude of some trends. However, they did not overestimate them, but underestimate.'

Even predictions that up to now have had little reliability, such as the loss of the rain forest or thawing permafrost, turn out to be disastrous realities. The tipping points of Earth's climate are coming closer – and afterwards there will be no chance of survival on this planet as *Homo sapiens*. Stephen Hawking would be right in his suggestion that, 100 years from now, humans will have disappeared from the planet. Rahmstorf explains the idea of the tipping point simply: 'Imagine pushing a cup full of coffee over the edge of your desk. For the first centimetres little happens; the cup just changes its position. But eventually the cup will tilt, fall and pour its contents onto the carpet.' There is no way back for the coffee.

Says Eckart von Hirschhausen, medical practitioner and comedian: 'The Earth has a serious infection with a herd of the species *Homo sapiens.*'

2. Four hundred million climate refugees

According to the World Bank, by 2030 about 100 million to 140 million climate refugees will wander the planet looking for watering holes. And the UN expects more than 400 million climate refugees by 2050 if we do not stop global warming. The climate crisis will thus become a crisis of democracy and society that ultimately brings the threat of war.

There is not much time left. What do we have to learn to overcome the climate crisis?

Our biggest problem is not too many humans, but a lack of humanity.

3. Millions take to the streets with Greta

What a crazy world we live in! A sixteen-year-old girl holds up the climate mirror to us and what do we see? Ourselves! The Dalai Lama says: 'The belief in rebirth calls for more environmental and climate protection. Because we will come back to this planet and, for that reason, we want a good climate and a healthy Earth.' What a reminder! On 15 March 2019, Greta, in her simple but determined language ('Our house is on fire', 'I want you to panic') drove 1.6 million young people onto the street. She went on to initiate the first worldwide climate strike on 20 September 2019, with more than six million protesters.

Greta is like the child in Hans Christian Andersen's fairy tale 'The Emperor's New Clothes' who calls out to the unseeing adults: 'But he has nothing on at all'. Greta calls out, 'No life without a good climate.' The world begins to wake up, rubs its eyes and realises that the child is right.

These days, the children and young people who demonstrate to support the climate behave more like grown-ups than the adults. When the German speaker of the Fridays-for-Future movement, geography student Luisa Neubauer, spoke to the shareholders of the energy supplier RWE, her microphone was taken away. What fear of the truth!

A week later, millions of mainly young people took to the streets – in Wellington and Vienna, in

Stockholm and New Delhi – to demonstrate with Greta for better climate protection. In Italy alone there were more than a million protestors. In Naples, placards read, 'We want hot pizza, not a hot planet'.

After huge demonstrations in Germany, the Greta effect also reached the political parties: surveys showed that the Greens, with 27 percent, were neck and neck with the CDU/CSU alliance (the Christian Democratic Union of Germany and Christian Social Union in Bavaria), once advocates of nuclear power. The SPD, the only party advocating coal, just reached 13 percent.

4. What could rescue look like?

First: by 2035 at the latest, we will reach 100 percent renewable energy. This is no problem, because the Sun sends us 15,000 times more energy than we are currently consuming. Don't think only of solar heat; think also wind energy, hydropower, biomass and geothermal energy – they are all ultimately powered by the Sun. This energy is environmentally friendly, forever and free. The Sun and the wind don't send out any bills. What an incredible economic advantage for future ecological energy supply. The insight is simple and irrefutable – like one of Greta's observations – so why do 90 percent of all roofs in Germany still stand around uselessly without solar panels?

Everyone learns at school that everything revolves around the Sun. So why don't we see the light? We must finally open up to energy from above. Without the Sun, there's no life.

Second: rapid withdrawal from coal and a price limit on CO_2. Greta thinks this should be possible before 2038. The transition to renewable energy will create far more jobs than will be lost by the abandonment of coal. In any case, we have no choice. French President Emmanuel Macron agrees with Greta, saying: 'Our house is on fire.' Yet instead of extinguishing the flames, we are discussing the price of the firefighters' water. When do we finally call the fire brigade?

Third: rapid entry into electric mobility (whether e-bikes or electric cars) and a doubling of public transport. China, Norway, California and the Netherlands are showing the way. There will be distancing problems on public transport after the Covid pandemic, but they can be solved.

Fourth: a switch to organic farming. More and more municipalities renounce the use of pesticides and glyphosate, and individuals can take a similar stand. Our consumer behaviour is politics with the shopping trolley. What is at stake when you choose your food, ultimately, is elementary: fertile soil, drinking water, clean air, a moderate climate and forests that are good for our souls, which are under stress, and which store excess CO_2.

Fifth: worldwide reforestation and greening of the deserts. A study carried out by Zurich University, ETH, has recently shown that reforestation in the United States, Russia, China, Brazil and Canada could make up for more than two thirds of manmade CO_2 emissions. What are we waiting for?

The child and youth organisation Plant for the Planet, which was started in Starnberg, Germany, by eight-year-old Felix Finkbeiner, has shown the way. In the past 15 years, it has planted more than 13 billion trees worldwide. Children and young people are already playing a major role in ensuring a good future. We adults must finally learn to take the justifiable fears of our children and grandchildren seriously.

5. Dare for a future

Politicians and parties still have time to campaign for the survival programme I've just spelled out. Here's the motto they should use: the climate catastrophe endangers our well-being and our lives. Smart climate politics secures and preserves our well-being and guarantees the future of our children and grand-children. We must dare for a future.

People's courage has always been decisive, as in the French Revolution in 1789, the Suffragette movement in the early 1900s or the peaceful revolution in Germany in 1989. Often they began with one or two individuals who dared big. Like Greta, they were all alone.

Time is running out. The energy transition is expensive, true, but no energy transition will cost an even higher price, says German politician Wolfgang Schäuble: the future of humankind. It is not a question of sacrifice or renunciation, but rather a question of whether we desire any kind of future on this unique and wonderful Earth.

In this book, the Dalai Lama shows how alternative forms of economic activity can be developed. He describes how a Buddhist economy based on the common good will help; the principles of mindful-ness, nonviolence, compassion and modesty would benefit our personal as well as our political and economic behaviour.

Is this book reason for alarmism? No way! It is a declaration of love for the future.

3

Save the Environment – The Dalai Lama's Climate Appeal to the World

1. Buddha would be Green – Me too, I am Green
Buddha was born as his mother leaned against a tree for support. He attained Enlightenment seated beneath a tree, and passed away as trees stood witness overhead. Were Buddha to return to our world, therefore, he would certainly be connected to the campaign to protect the environment.

Speaking for myself, I have no hesitation in supporting initiatives related to environmental protection, because threats to our environment imperil our very survival. This beautiful blue planet is our only home. It provides a habitat for unique and diverse communities. Taking care of our planet is to look after our own home.

True, if we compare environmental damage to warfare and violence, it's clear that violence has a more immediate impact. The trouble is that damage to the environment takes place more stealthily, so we often don't see it until it is too late. We have reached a tipping point in global warming.

2. Environmental education

Environmental education about the consequences of the destruction of our ecosystem and the dramatic decrease in biodiversity must be given top priority. But creating awareness is not sufficient; we must find ways to bring about changes in the way we live. I call on the younger generation: be rebels in demanding climate protection and climate justice because it is your future that is at stake.

One of the most positive recent developments has been the growing awareness that we have to act. Sixteen-year-old Greta Thunberg, the teen-age environmental activist who insists we heed scientists' warnings and take direct action, inspires me. Millions of young people have been moved by her example to protest governments' inaction over the climate crisis. She is correct to say, 'No one is too small to make a difference.' I wholeheartedly support Fridays for Future, the movement she initiated.

I am encouraged to see young people's determination to bring about positive change. They are confident of making a difference, because their efforts are based on evidence and reason.

More and more people understand that the survival of humanity is at stake. Simply meditating or praying for change is not enough. There has to be action.

3. Universal responsibility

It is no longer enough to think only of 'my country', 'my people', 'us' and 'them'. We must all learn to work for the benefit of all human beings.

Humans are social animals born with a sense of belonging to a community. We have to realise that, just as our future depends on others, theirs depends on us. Our world is deeply interdependent, not only in terms of our economies but also in facing the challenge of climate change.

We have to appreciate that local problems have global ramifications from the moment they begin. The climate crisis affects the whole of humanity.

Island states such as Fiji, the Marshall Islands, the Maldives and the Bahamas have shown that collectively we can make a difference. The 2015 Paris Agreement to combat climate change, which was signed by 196 countries, was a good start, but it has to be followed through with action.

We need a sense of universal responsibility as our central motivation to rebalance our relations with the environment and with our neighbours. Appreciating the oneness of humanity in the face of the challenge of global warming is the real key to our survival.

4. The revolution of compassion

I am now 84 years old and have lived through many of the upheavals of the 20th century: the destruction and suffering brought by war, as well as unprecedented damage to the natural environment. Today's younger generation have the ability and opportunity to create a more compassionate world. I urge them to make this 21st century an era of change rooted in dialogue and a century of compassion for all the inhabitants of this planet.

Over-exploitation of our natural resources results from ignorance and greed, and a lack of respect for life on Earth. Saving the world from the climate crisis is our common responsibility. We must find ways to exercise freedom with responsibility.

We need a revolution of compassion based on warm-heartedness that will contribute to a more compassionate world with a sense of oneness of humanity. The entire human family must unite and cooperate to protect our common home. I hope that efforts to achieve a more sustainable way of life will meet with success.

4

Franz Alt's Interview with His Holiness, the Dalai Lama

1. The purpose of life is to be happy
Franz Alt: Your Holiness, dear friend. Fifteen years ago you said to me in an interview: 'The 21st century could become the happiest and most peaceful one in human history. I hope so for the youth.' Do you still cherish that hope?

Dalai Lama: I am hopeful that the 21st century could become the most important century in human history. The 20th century experienced immense destructions, human sufferings and unprecedented environmental damages. The challenge before us, therefore, is to make the 21st century a century of dialogue and promotion of the sense of oneness of humanity.

As a Buddhist monk, I appeal to all human beings to practice compassion, which is the source of happiness. Our survival depends on hope. Hope means something good. I believe the purpose of life is to be happy.

The world's seven billion human beings must learn to work together. This is no longer a time to think only of 'my nation' or 'our continent' alone. There is a real need for a greater sense of global responsibility.

I feel optimistic about the future because humanity seems to be growing more mature; scientists are paying more attention to our inner values, the training of the mind and the emotions. There is a clear desire for peace and concern for the environment.

Franz Alt: The Paris Climate Summit at the end of 2015 was the beginning of a new reality. For the first time, the world may have seen itself as a world family. All governments worldwide and the European Union committed themselves in writing not to increase global warming by more than 1.5 degrees Celsius, 2 degrees at the most, compared with 1980 levels. But globally, we already have an increase of more than 1 degree. If we continue like that, global warming could rise by 5 or 6 degrees, even in this century. I would not want to be my grandson then. As the German proverb says, paper does not blush — but governments are not acting. Are you still optimistic? Can the Paris Agreement still be achieved?

Dalai Lama: I hope and pray that the 2015 Paris Agreement will finally bring tangible results. Egotism, nationalism and violence are fundamentally wrong.

The United States's withdrawal from the Paris Agreement is very sad. It is important for scientists to continuously speak up about the dangers we face and alert the public. Here, the media has an important responsibility in educating the people. The gap between rich and poor is also very serious and we have to take steps to close it by helping the poor.

Any human activity should be carried out with a sense of responsibility, commitment and discipline. But if our activities are carried out with short-sightedness and for short-term gains like money or power, then they all become negative and destructive activities. Protecting our environment is not a luxury we can choose to enjoy but a matter of survival.

Even a small rise in our body's temperature can cause us discomfort, while an increase of five to six degrees can be life-threatening. Annually, we have been witnessing global warming due to climate change. Recently, both America and Europe have experienced extremely hot summers and cold winters. The questions of the environment and climate change are global issues, not just concerns for Europe, Asia, Africa or the Americas. What happens on this blue planet affects us all.

It is not sufficient to just express views and hold conferences. We must set a timetable for change.

Franz Alt: As early as 1992 you said: 'Universal responsibility is the key to human survival'. What does that mean in concrete and practical terms?

Dalai Lama: Human beings are social animals and must learn to live together. This is no longer a time to think only of 'my country', 'my people', 'us' and 'them'. We live in a global world, but countries think about their own national interests rather than the world's interests. That needs to change, because the environment is a global issue. In order to protect global environmental issues, some sacrifice of national interests is needed.

2. We are all children of one world

Franz Alt: Nationalism has been shaping our history for centuries. Is it really possible to overcome nationalist thinking?

Dalai Lama: Wherever I go, I emphasise that all seven billion human beings are physically, mentally and emotionally the same. Everybody wants to live a happy life free from problems. Even insects, birds and animals want to be happy.

In order to ensure a more peaceful world and a healthier environment, we sometimes point a finger at others, saying they should do this or that. But change must start with us as individuals. If a single individual becomes more compassionate, it will influence others, and so we will begin to change the world. Scientists say our basic nature is compassionate. This is very positive.

In the face of such global problems as the greenhouse effect and depletion of the ozone layer caused by chlorofluorocarbon (CFC) gases, individual organisations and nations are helpless. When I was awarded the Nobel Peace Prize in Oslo in 1989, I called upon the world to assume universal responsibility. We have to learn that we are all brothers and sisters who live on one Earth, under the same Sun.

Unless we all work together, no solution can be found. Therefore, our key responsibility is to commit ourselves to the ethical principles of universal responsibility beyond profit and religion, and to place the well-being of all sentient beings and future generations above our egoism. Climate change is an issue that affects the whole of humanity. But if we have a genuine sense of universal responsibility as our central motivation, then our relations with the environment will be well balanced, and so will our relations with our neighbours. Our Mother Earth is teaching us a lesson in universal responsibility.

Therefore, each of us as individuals has a responsibility to ensure that the world will be safe for future generations, for our grandchildren and great-grandchildren.

Franz Alt: Is global warming just a political problem or can every individual do something about it?

Dalai Lama: According to scientists, human beings are responsible for global warming and the change in weather conditions. Logically, this means that we also have a responsibility to solve the problems that we have created.

On an individual level, we should change our lifestyles, consume less water and electricity, plant trees and reduce the use of fossil fuels, which took millions of years to form. Fossil fuels are non-reusable energy;

we must instead use renewable energy, such as solar, wind and geothermal.

As a boy studying Buddhism, I was taught the importance of a caring attitude towards the environment. The Buddhist practice of nonviolence applies not just to human beings but to all sentient beings.

What distinguishes human beings from animals? It is our specific capacity for long-term thinking. Animals live only from one day to the next, whereas our brains can think ten or even a hundred years ahead. A thousand years may be too much for us. In consequence, we are equipped to make preparations for the future and plan for the long term.

3. Without humans the Earth would be doing better

Franz Alt: Is it only our shortsightedness that prevents us from treating our natural environment carefully?

Destruction of nature and its resources results from ignorance, greed and lack of respect for the Earth's living things. Today, we have access to more information, and it is essential that we re-examine ethically what we have inherited, what we are responsible for and what we will pass on to the coming generations.

Resolving the environmental crisis is not just a question of ethics but a question of our own survival. The natural environment is very important not only for those of us alive now but also for future generations. If we exploit it in extreme ways, even though we may get money or other benefits from it now, in the long run we ourselves and future generations will suffer. When the environment changes, climatic conditions also change. When they change dramatically, the economy and many other things change as well. Even our physical health can be greatly affected.

Franz Alt: In the past, humans needed protection from their environment. Today, it is the other way round. Scientists tell us that without humans the Earth would be doing better.

Dalai Lama: As someone born in Tibet, the roof of the world, where the world's highest peaks are to be found and Asia's great rivers originate, I have loved nature since my childhood. I have made environmental conservation one of my life's commitments and advocate protection of the environment wherever I go. Therefore, I call on everyone to speak out about global warming, which affects the future.

Franz Alt: In his Encyclical on the environment, Pope Francis puts it like this: 'The interdependence of all creatures is God-given. The Sun and the Moon, the cedar and the field flower, the eagle and the sparrow – the myriad of differences and inequalities is evidence that creatures are not self-sufficient, but exist only in dependence on each other and complement one another in mutual service.' Does the Pope's statement match with what you think?

Dalai Lama: I welcome Pope Francis's Encyclical on the environment. I also see similarities between the Pope's Encyclical 'One human family, one common home' and my message of the oneness of humanity. Since global warming and climate change affect us all, we have to develop a sense of the oneness of humanity and universal responsibility.

The metaphysics of the wise men of ancient India and the West are converging in times of ecological crisis. Technology alone will not save us. We need interdependence of ethics and technology. We need a joint plan to save the planet.

Caring for the Earth is our shared responsibility. Each one of us has a moral responsibility to act, as so powerfully stated by the Pope's Encyclical.

Franz Alt: In our previous book, The Way to Peace in a Time of Division, *you expressed the idea that 'Ethics is more important than religion'. What does that mean as far as environmental policy is concerned?*

Dalai Lama: Religion should not be just limited to praying. Ethical action is more important than prayers. What is Buddha, Allah or Christ supposed to do if we human beings destroy our Earth; fill the oceans with plastic so that fish, seals and whales perish; cause a rapid increase of desertification; and release greenhouse gases into the atmosphere? Christ, Allah or Buddha are not responsible for climate change and the destruction of the environment; it is a human-made problem. Therefore, we must take responsibility and find solutions to the problems. That is why we need environmental ethics that focus on action and compassion for all sentient beings.

Scientists have concluded that basic human nature is compassionate. Those who grow up in a more compassionate atmosphere tend to be happier and more successful. On the other hand, scientists suggest that living with constant anger or fear undermines our immune system. Hence, compassion and warm-heartedness are not only important at the beginning of our lives but also in the middle and at the end. Their necessity and value are not limited to any specific time, place, society or culture.

4. The Himalayan glaciers are vanishing

Franz Alt: Two thirds of the Himalayan glaciers are in danger of disappearing by 2050 due to global warming. That would affect the water supply of billions of people in India and China.

The Bible states: 'In compassion, righteousness and peace will kiss each other.' And in the New Testament, Jesus says, 'Be compassionate even as your heavenly Father is compassionate'. Our actions are only compassionate when they result from our solidarity. In North Africa I have seen areas that were a paradise for humans yesterday that today are already drought-stricken; tomorrow they will be uninhabitable. Over 50 years, I have observed similar catastrophic developments in India and Bangladesh. Can ice melting and global warming be stopped?

Dalai Lama: Billions of dollars are spent on weapons of mass destruction. If half that sum was used to develop new technologies and wider use of renewable energy, the positive impact it would have in our efforts to limit global warming would be tremendous!

Putting our hope in the younger generation is not sufficient. Politicians, too, must act urgently. It is not enough to hold meetings and conferences. We must set a timetable for change. Only if they start to act now will we have reason to hope. We must not sacrifice our civilisation for the greed of the few. Journalists have an equally important role. I tell them that in this crisis they have a special responsibility to bring awareness to the people – not just to report on bad news, but also to bring people hope.

Recent studies suggest that the world is getting close to exceeding its carbon budget. Therefore, this budget must become the most important currency of our time. Politicians are gradually running out of excuses, but we must use our time wisely.

Hundreds of thousands of young people are taking to the streets in the new global youth environmental movement, Fridays for Future, to persuade politicians towards better climate protection. I am encouraged to see young people's determination to create greater awareness to bring about positive change. They will succeed because their efforts are based on scientific truth and reason.

This little book is a call to action! An appeal to all politicians, opinion formers, journalists, religious leaders – to all people. Since the future of all coming generations rests on our shoulders, we must be determined to take action before it is too late.

Franz Alt: What you have just said about politics and politicians also applies to we journalists. We must finally begin to describe the crisis as a crisis. We are facing unimaginable suffering for billions of people. Why is the climate issue a matter of survival for all life?

Dalai Lama: I often joke that the Moon and stars look beautiful, but if any of us tried to live on them we would be miserable. This blue planet of ours is a beautiful habitat. Its life is our life, its future our future. Indeed, the Earth acts like a mother to us all. Like children, we are dependent on her. Our world is deeply interdependent, in terms of both our economies and the problems like climate change that challenge us all.

Scientists say even a small rise in Earth's temperature is a risk to human beings, animals, agriculture, water; it causes the melting of glaciers in the Arctic and Antarctic, in Greenland and Alaska, in the Himalayas and in the Alps. If the world fails to halt global warming, small island states may disappear forever due to rising sea levels. Unfortunately, the poor are hit hardest in weather-related disasters.

When we see photographs of the Earth from space, we see no boundaries between us, just this beautiful blue planet. There is a real need for a greater sense of global responsibility based on the oneness of humanity.

Franz Alt: Why do you say that Tibet is the epicentre of climate change?

Dalai Lama: A Chinese ecologist has described the Tibetan Plateau as the Third Pole because it is the third largest area of frozen water on the planet, after the North and South Poles. The effects of global warming on the Tibetan plateau have a significant impact on the lives of more than 1.5 billion people living in the region. The same article mentioned that the temperatures on the Tibetan plateau have increased by 1.5 degrees Celsius, more than double the global average. The Third Pole's glaciers are melting at a rate that has almost doubled since 2005. More than 500 small glaciers have disappeared altogether, and the biggest ones are shrinking rapidly, according to the research.

The Tibetan plateau happens to be the largest water tank in the world. All the major rivers of Asia – including the Ganges, Karnali, Brahmaputra, Indus, Sutlej, Irrawaddy, Salween, Yellow River, Yangtse and Mekong – originate there. More than 1.5 billion people live by these waters: one fifth of the world's population. Without water there's no life. If Tibet's 46,000 glaciers continue to melt, we will face unimaginable water problems and water will probably become a key cause for conflict in the future. So the ecology of Tibet is something really important.

Franz Alt: We were both standing at the Berlin Wall in November 1989. And in East and West Berlin the Mauerspechte *– 'wall-peckers' – were already hammering at this inhuman monster. People on either side of the wall gave you a burning candle and lifted you onto the remains of the wall. There you said these powerful words: 'As sure as this wall will fall, my home, Tibet, will one day have freedom.' For me, an unforgettable moment. After all, only a few months before, China had brutally ended the student revolt on Tiananmen Square. Afterwards, we discussed your optimistic perspective in* *front of several thousand students at the Freie Universität Berlin. Would you repeat that sentence today, given that the repression in Tibet under Chinese occupation has increased since 1989? Do you remain such an optimist?*

Dalai Lama: When we came into exile, preserving our identity, language and culture was our priority. These days, when Tibetans in Tibet reveal their passion likewise to preserve their culture, Chinese hardliners oppose it and claim it is indicative of their 'separationist' motives. Despite restrictions, Tibetans' spirit remains resolute.

Things are changing and a totalitarian system has no future.

5. A nuclear war would be the last in the history of humankind

Franz Alt: You fear wars over water between India and China. Both countries have nuclear bombs. Could there ever be a nuclear war between India and China on the water issue?

Dalai Lama: A nuclear war would probably be the last conflict in human history, because nobody would be left to wage another.

Franz Alt: Why is it important to maintain harmony between the natural environment and the sentient beings?

Dalai Lama: There is a very close interdependence between the natural environment and sentient beings living in it. Thus, we share a sense of universal responsibility for both humankind and nature.

When the environment becomes damaged and polluted, there are many negative consequences. Oceans and lakes lose their cool and soothing qualities, so the creatures depending on them are disturbed. The decline of vegetation and forest cover causes the Earth's bounty to decline. Rain no longer falls when required, the soil dries and erodes, forest fires rage and unprecedented storms arise. We all suffer the consequences.

Before the Chinese occupation, Tibet was a fresh, beautiful, unspoiled wilderness sanctuary in a unique, natural environment. Sadly, in the last six decades, Tibetan wildlife and the fragile ecology have been almost destroyed by the Chinese occupation. What little that is left must be protected. Every effort must be made to restore the Tibetan environment to its balanced state.

In spite of all the suffering that China has inflicted on Tibetans for over six decades, I remain convinced that most human conflicts can be solved through sincere dialogue held in a spirit of openness and reconciliation. We have learned that even enemies can become friends. I am a strong believer in nonviolence.

Franz Alt: You have said, 'The environmental catastrophes are the reflection of our combative and destructive ways of thinking based on a selfish desire for prosperity and profit.' Isn't a certain egoism and pursuit of wealth part of humans' essence?

[After a moment's reflection and a smile, the Dalai Lama answers.]

Dalai Lama: Material values do matter. But deeper inner values are more important than material values. In the last century, we made great material progress. But it is precisely this material progress that is now leading to environmental destruction. We need a new balance between economy and ecology, otherwise we will destroy the basis of our lives. Material progress alone cannot reduce our psychic stress, anxiety, anger and frustration.

My friend Mikhail Gorbachev, former president of the Soviet Union, is still committed to environmental issues with the international organisation Green Cross, which he co-founded. Ecology must become the smarter economy. Only then will we be able to live in a sustainable way.

As to whether the world is getting better or worse, there is growing opposition to the existence of nuclear weapons. Where no one used to talk about the environment, today it is on everyone's lips. Scientists who once only paid attention to material things are now paying attention to training the mind. I am optimistic that people are generally becoming more mature.

I already described us as selfish, that's right. But we should be wise-selfish rather than foolish-selfish. Think less about 'I' and think more about others' well-being. You get maximum benefit. So that is wise-selfishness.

Franz Alt: You talk about prioritising environmental education from kindergarten to secondary schools, but also in universities. Why is it important to start so early in children's education?

Dalai Lama: Every child should learn in school that his or her own future and happiness will always depend on the future and happiness of others. Even in kindergarten, children can learn that all seven billion human beings have the right to be happy. We all live on the same planet, under the same Sun, and breathe the same air. Today, the world needs environmental ethics education based on a deeper understanding that transcends religion. Children can learn to appreciate this at school.

Environmental education must be given top priority, as we all have become witnesses to the destruction of our ecosystem and a dramatic decrease in biodiversity. Environmental education means learning to maintain a balanced way of life. In order to maintain a universal appeal, such ethics need to have a secular basis.

When I came from Tibet to India in 1959, I had no idea of the problems with the environment. When I first heard 'You cannot drink this water', I was surprised that it was polluted. In Tibet, passing by a stream, we always would enjoy the waters. No problem. I learned about pollution and gradually about ecology. I now feel a deep concern about the environment, as it has become a question of our survival. I learned about this not through meditation but through awareness, with help from experts.

We may talk about going to the Moon or to Mars, but we can't settle there. This Earth is the only place where we can live.

6. More education of the heart

Franz Alt: What do you mean by the phrase 'education of the heart'?

Dalai Lama: My wish is that more attention is paid to educating the heart: teaching love, kindness, peace, compassion, forgiveness, mindfulness, self-discipline, generosity and tolerance. This education is necessary from kindergarten to secondary schools and universities. By education, I mean social, emotional and ethical learning. We need a worldwide initiative for educating the heart and training the mind in this modern age.

At present, our modern educational systems are oriented mainly towards material development. Modern education is not adequate. It pays little attention to inner values, yet our basic human nature is compassionate. Therefore, we need to develop a curriculum in the modern education system based on compassion and warm-heartedness to make it more holistic.

We are not like plants; we have emotions. We need to learn to manage our emotions and achieve inner peace. Our education should include an understanding of how to achieve peace of mind. Education should teach us how to live properly, how to balance our wish for physical comfort with mental comfort. That's what's important.

We must learn that humanity is one big family. We are all brothers and sisters: physically, mentally and emotionally we are the same. But we are still focusing far too much on our differences instead of our commonalities. After all, every one of us is born the same way and dies the same way.

5

The Solar Age Begins –
The Sun Wins

1. Solar energy is social energy

Franz Alt: In our discussions, we have often talked about solar energy. The Sun sends 15,000 times more energy to Earth than we currently consume. And this power is free, environmentally friendly, worldwide and eternal. We can make use of the entire symphony of renewable energies. In some countries, such as Costa Rica or Iceland, the entire energy industry is already renewable. Worldwide, one third of electricity production is renewable. There is no lack of knowledge or technologies, only of rapid implementation. Why does transition take so long?

Dalai Lama: You told me that Germany today produces as much as 50 percent green electricity. In 2000, it was only 5 percent. This shows even industrialised countries with high energy consumption can switch to renewable energy. I understand that storage technologies for solar and wind power are now more developed. Besides, the Sun and wind do not send an invoice. They are nature's gifts, which we should use much more in the future. Solar and wind are already more economical sources of energy worldwide, so we don't need nuclear or coal-fired power plants. We are at the beginning of a world-wide solar revolution.

We must change our lifestyle and our reliance on old energy. So, there must be more government incentives for renewable energy companies and for ordinary people to use renewable energies.

Franz Alt: May I ask, dear friend: we can already build houses and factories that produce more green energy than they use. The Sun shines on every roof. We know that, in the future, ships and aeroplanes will use solar-generated hydrogen. In the past 20 years, we have created more than 11 million sustainable jobs worldwide through renewable energy. The International Renewable Energy Agency estimates that the solar energy transition will create 25 million new jobs by 2030. But why is the energy transition so slow?

Dalai Lama: New technology has always taken a relatively long time to make a complete breakthrough. More and more companies are producing electric cars, but if the car is too expensive, only the very rich can buy them. So these cars must be more affordable. Similarly, other renewable energy must be made more affordable, especially to the poorer in society, who are most vulnerable to climate change. I have learned from scientists that the use of solar power and wind power worldwide has increased over the years. We are making good progress. As I said earlier, there must be more education and incentives for the use of renewable energies.

2. We should lock up the politicians

Franz Alt: We can already produce solar power in Africa or in Chile at 2.5 cents per kilowatt-hour. The government of Saudi Arabia wants to produce solar power at 1 cent by 2025 in the largest solar power plant in the world. Renewable energy is the decisive step to prosperity for all. With cheap energy, the economies can be developed in the poor countries of the Southern Hemisphere, eliminating the poverty that causes migrants to leave. How can we get politicians to swiftly implement what they decided upon in Paris?

Dalai Lama: [Breaking into his world-famous laugh and grinning] Perhaps you should lock up the world's most important politicians in a room for a while and pipe carbon dioxide into it until they realise what climate change really means. Then, they would probably very soon feel what greenhouse gases are doing to us humans [still laughing quite loudly].

I very much appreciate the worldwide activities for energy transition and for more environmental protection. Because I often get the impression that politicians do not take climate and environmental protection seriously enough. Ignorance is the number-one enemy.

Scientists say that, due to global warming, many parts of the world could become desert. This is very serious.

Dr Yuan T. Lee, a Taiwanese Nobel laureate in Chemistry, told me that in another 80 years the world will be like a desert. He said that water resources are already diminishing alarmingly. Therefore, we all need to modify our lifestyle, abandoning fossil fuels and turning to renewable sources of energy.

One of my dreams, perhaps an impossible dream, is to harness the solar potential of places like the

Sahara Desert and to use the power to run desalination plants. The fresh water thus produced could green the desert and produce food crops. It is a project that would have widespread benefits but it would function on a scale that requires global cooperation.

So, taking care of the environment, taking the necessary steps to reduce global warming, is a serious matter. I'm a monk so I have no children, but people who have children have to think about how life will be for them and their grandchildren. We're at the start of the 21st century, but we should also be looking ahead to how things might be in the 22nd and 23rd centuries.

3. Rebirth calls for environmental protection

Franz Alt: Could a consciousness of reincarnation also help? If someone knows that he or she will come back, they want a healthy planet. Rebirth is taken for granted in Asian religions. We also know that Jesus spoke repeatedly of rebirth, and the Jewish philosopher Shalom Ben-Chorim says a belief in rebirth was popular at the time. Only later did the early Christian bishops eliminate Jesus's words. But in all cultures, there are people who remember past lives. Religions that believe in rebirth should, if only for selfish reasons, stand up for environmental protection, because in their next lives they also need a hospitable Earth. Does environmental and climate protection have a higher priority in Asian rebirth religions than in Western religions?

Dalai Lama: Yes it does. But hardly in practical politics. In practical environmental behaviour, very little can be felt so far. Think of China's environmental problems. Or of Japan's nuclear power plants. Or coal-fired power stations in India.

It is true, however, that a person who believes in rebirth naturally wants an environmentally friendly planet in their next life. Me too [laughing].

We can no longer keep exploiting the resources of this earth – the trees, the water and the minerals – without any care for the coming generations. It is common sense that we cannot survive if we keep working against nature. We must learn to live in harmony with nature. As a Buddhist monk who believes in rebirth, even for selfish reasons, I believe that we must pay more attention to our planet. Because we will come back. And all of us would like to live on a healthy Earth. The belief in rebirth calls for more environmental and climate protection.

Franz Alt: Wise men of all cultures have been convinced of reincarnation, from Pythagoras in ancient Greece decades before Buddha to the German philosopher Arthur Schopenhauer in the 19th century and the Christian church father Origen of Alexandria. The Christian West is the only region where reincarnation is officially denied. You, dear friend, say, 'Spirituality is the essential key to our survival'. Could you give reasons for this statement?

Dalai Lama: I have often said that in keeping with the tradition of Tibetan Buddhist culture: all sentient beings have been our mothers. All of Buddhist spirituality is characterised by this realisation. All sentient beings are connected by a maternal bond. This is the basic truth of awakening, enlightenment and realisation. We are all interconnected in the universe and, from this, universal responsibility arises. Jesus knew this spiritual law, which is called 'karma' in Buddhism, and talked about it without using the word *karma*. It is a spiritual law that 'you reap what you sow'. Things depend entirely on your effort, your action. So things change through action and not by prayer. We must act to create positive karma. Positive karma means positive action.

4. Buddha: 'We are what we think'

Franz Alt: Buddha said, 'We are what we think. Everything we are arises from our thoughts. Our thoughts shape our world.' In the recent past, Greta Thunberg has shown what a single person can achieve. She started her school strike in the summer of 2018 and on a Friday sat down all by herself in front of the Reichstag in Stockholm. On her protest poster it said 'School strike for the climate'. On the following Friday, four girls and boys joined her. Soon, hundreds of thousands followed her in over a hundred countries, and on 15 March 2019, it was 1.6 million. On 20 September 2019, more than six million.

The young lady spoke at the World Climate Summit in Poland, met the Pope, was nominated for the Nobel Peace Prize and in Sweden was named 'Woman of the Year'. Time Magazine *counts her among the 100 most influential people worldwide. When asked why she is fighting for the climate, she says: 'I know what's at stake – the survival of mankind. And I consider it my moral duty to do everything I can to avert the worst. At first, I tried to inspire others, but nobody wanted to get involved, so I started alone. We can achieve a lot, if a lot of people join in.' She also says that she took part in* *many demonstrations for the climate before, but nobody reported on it. Only when she had the idea of a school strike did it become a world issue. And today she tells the politicians: 'We're going on strike until you act. Together we will change the world.' The shy sixteen-year-old girl fell ill because she could no longer stand the images of plastic mountains in the oceans. Her mother says: 'After her illness, Greta sees things that others don't see: the CO_2 of aeroplanes and cars and coal-fired power stations. She sees that we are transforming the atmosphere into an invisible gigantic garbage dump.' As a Buddhist, how do you explain the global 'Greta effect'?*

Dalai Lama: I really appreciate Greta Thunberg's efforts to raise awareness of the need to take direct action. People like her are realistic. We should encourage them.

Greta's motivation to create awareness of global warming among school children is a remarkable achievement. Despite being very young, her sense of universal responsibility to act is wonderful. I support her Fridays for Future movement.

I believe that every individual has a responsibility to help guide our global family in the right direction.

Good wishes alone are not enough; we have to take responsibility. Large human movements spring from individual human initiatives.

The youth of the 21st century are the planet's real humanity now. They have the ability and opportunity to bring change, to create a century of peace, dialogue and compassion. Even as global warming increases in intensity, they can work together in the spirit of brotherhood and sisterhood to share and find solutions. They are our real hope.

Ideas may travel from the top down, but the movements that put them into effect have to work from the bottom up. Therefore, I am encouraged to see young people trying to bring about positive change. I am also confident, because their efforts are based on truth and reason – therefore, they will succeed.

So now the generation of the twenty-first century, the young brothers and sisters, must take a more active role in protecting ecology and our home.

5. Greta: 'Our house is on fire'

Franz Alt: Are Greta and her young followers right when they call out to we older people: 'We are loud because you're stealing our future'? Greta says: 'Our house is on fire.' Is that exaggerated?

Dalai Lama: The young climate activist is right. Scientists and environmental campaigners have been selflessly and tirelessly putting effort into creating a better environment for the world so that future generations will be able to live a healthy, happy life. The 2015 Paris Agreement was signed by leaders of 196 countries to combat climate change and limit temperature rise to 'well below' 2 degrees Celsius, and it is a source of hope and encouragement. Governments are now committed to more effective climate protection. If millions of young brothers and sisters all over the world go on strike because politicians are inactive, that's a sign that something is not working.

Climate change is not the concern of just one or two nations. It is an issue that affects all humanity. This beautiful planet is our only home. If, due to global warming or other environmental problems, the Earth cannot sustain itself, there is nowhere else we can move to. We have to take serious action now to protect our environment and find constructive solutions to global warming.

Franz Alt: How can we motivate politicians as well as businesspeople to do more for the environment and the climate than they have done so far?

Dalai Lama: Before the pandemic struck, millions of young brothers and sisters were protesting, calling for politicians to take action on combating climate change.

We must once more think globally, but act locally. This should even apply when electing political leaders. Our voting patterns are also an ethical issue. Today we are witnessing a strong connection between environmental politics and elections.

People have elected greater numbers of Green parliamentarians in Germany, Switzerland, Finland, Belgium, the Netherlands and the European Parliament. This is a strong indication that public opinion and actions can change politicians' minds.

Fortunately, young people today understand the connection between environmental politics and elections.

Franz Alt: What's your response to what Greta Thunberg and young people like her are doing? Do you support young students taking time off from school and standing up as activists to demand radical change?

Dalai Lama: I wrote a letter to her. I really admire what she's doing. We older people will probably be able to manage over the next one or two decades. But the lives of young people like her may extend until the end of this century, and they will have to face whatever changes come. Therefore, it is quite right that students and today's younger generation should have serious concerns about the climate crisis and its effect on the environment. They are being very realistic. Sometimes it seems that older people lead a more materialistic way of life; they belong to a more materialistic culture. Younger people are beginning to feel there is something lacking in such a way of life. We should encourage them.

Franz Alt: Greta Thunberg is quite realistic about politics. She said to members of the US Congress: 'Don't invite us here to just tell us how inspiring we are without actually doing anything about it.' What can we do about climate change now?

Dalai Lama: Well, we could do a lot. You come from Germany. European history since 1945 has shown that peace is possible, even though in Europe in the last century everyone was at war against one another. I have great admiration for the spirit of the

European Union, which has preserved peace among its members. No country within the European Union has waged war against another. Seventy years of peace! The European Union was rightly awarded the Nobel Peace Prize in 2012. Politics can change just as people can. The European Union is a wonderful peace project that gives me great encouragement.

Every crisis creates opportunities. Many people experience this in their private lives. But crises in politics and the economy also always create opportunities. We are always the same people – at all levels.

6

The Mountains Here are as Bald as a Monk's Head

1. Plant trees

Franz Alt: As a Buddhist monk, you rely on the power of thought. Thoughts are energies that form in our minds, and their energies are transported further on a spiritual level. Our positive energies can have a positive impact, negative thoughts, of course, a negative effect. What could this Buddhist thinking do for more and better climate protection?

Looking at the Himalayas from your exile, I recall one of your quotations: 'The Himalayan mountains have become as bald as a monk's head.' [The Dalai Lama laughs and scratches his bald head.] Thirty years ago I showed on German television the brutal deforestation of Tibet by the Chinese. How can we stop the environmental destruction and how can the climate still be saved?

Dalai Lama: Only when we understand that our earth is like a mother, Mother Earth, will we really take care of her. We Tibetans, like the ancient Indian peoples, understand this interdependence: healthy Earth, healthy animals, healthy plants, healthy forests, healthy water, healthy people. Mother Earth warns us today, 'My children are behaving badly'; she is warning us that there are limits to our actions.

Today we are consuming huge amounts of coal, gas, oil and petrol every day that took nature a million years to form. That is the cause of global warming. As a Tibetan Buddhist monk, I am committed to a moderation of our consumption patterns. A responsible life is a simple and contented life. We must learn to cooperate, work and live with nature, not against it.

Franz Alt: In Tibet, China has cut down 85 percent of all trees, thus depriving your country of its life force. Why did the Chinese cut down Tibet's forests and what are the consequences for your native county?

Dalai Lama: When the forests in Tibet die, a whole nation suffers. And when a people suffers, the whole world suffers. We need forests for our health, too. When we go for a walk in a forest, fresh air is healing. We need green forests. They are nature's great gift. Forests are good for our souls. In the forest, we find the calm that our brains need for regeneration. Forests are water reservoirs, home to many animal and plant species, and are important as huge air-conditioning machines. They are a mirror of the diversity of life.

The large-scale deforestation in Tibet is a matter of great sadness. It is not only sad for the local area, which has lost its beauty, but for people in Tibet and elsewhere. The deforestation of the Tibetan plateau, according to experts, will change the amount of sunlight reflected from snow into space [forested areas absorb more solar radiation], and this will affect the monsoons not only in Tibet, but in all surrounding areas. Therefore, it becomes even more important to conserve Tibet's environment.

The environmental destruction in Tibet clearly shows that Chinese Communist ideology lacks what, in our Tibetan culture, is meant by interdependence or universal responsibility. This also surprises me because communists like to sing 'The International' [again laughing]. Today, no nation can solve its problems on its own.

Franz Alt: Interjection, dear friend! In the United States, President Trump rules according to the slogans 'America first' and 'Make America great again'. Is this motto still up to date in times of globalisation?

Dalai Lama: When the president says, 'America first', he makes his voters happy. I can understand that. But from a global point of view, the statement is not relevant. In today's world, everything is interconnected. America's future also depends on Europe and Europe's future on that of Asian countries and so on. The new reality means that everything is related to everything. The United States is the leading nation of the free world. That's why the US president should think more about global-level issues.

Franz Alt: Should a contemporary motto not be, 'Make the planet great again'?

Dalai Lama: Certainly! The United States is still very powerful. The motto of Americans' ancestors was peace, liberty and democracy. The totalitarian systems have no future. As a leading power, the United States should ally itself closely with Europe. I am an admirer of the European Union. The EU is an exemplary peace project. Unfortunately, President

Trump had announced US withdrawal from the Paris Agreement. He may have his reasons, but I do not support them. The EU should also become a role model in terms of climate protection. Each and every individual should become a climate protector. However, we will not reach this goal through egoism and nationalism, but through the promotion of a sense of oneness of humanity.

Franz Alt: You suggest planting trees for the future and for peace. Why is that so important?

Dalai Lama: Trees have been our companions through history and they remain important today. They purify the air for living beings to breathe. Their shade provides a refreshing place to rest and serves as a place for insects and birds to live. They contribute to timely rainfall, which nourishes crops and livestock, and balances the climate. They create an attractive landscape, pleasing to the eye and calming for the mind, and continually replenish their surroundings. Properly managed, they are also a source of economic prosperity.

When the environment becomes damaged and polluted, oceans and lakes lose their cool and soothing qualities, so the creatures depending on them are disturbed. The decline of vegetation and forest cover causes Earth's bounty to decline. Rain no longer falls when required, the soil dries and erodes and forest fires rage. We all suffer the consequences, whether we are ants in the jungle or human beings in cities.

In the context of Buddhism, trees are often mentioned in accounts of the principal events of Buddha's life. His mother leaned against a tree for support when Buddha was born. He attained Enlightenment seated beneath a tree, and trees stood witness overhead when he passed away. According to the monastic code of discipline, fully ordained monks are enjoined not only to avoid cutting trees or grass, but also to plant and nurture them.

Therefore, it is in our own interest to plant trees and flowers around the places where we live, work and study, as well as around hospitals and alongside paths and roads.

In the Tibetan monasteries in Tibet and India, we have been cultivating tree plantations over the past few decades. This brings in the action of serving others and of creating a better environment and a happier place. [Please also read the Dalai Lama's tree poem at the end of this book, p. XX] And to truly acquire a sense of responsibility for community, one has to first feel responsible for one's own place or home.

The longing for nature and greenery is ingrained in us. Human beings love greenery so much that they plant more and more trees in our cities and towns, and even trees on the roofs. When you spend time in the forest and hear birds singing, you feel good inside. The healing power of forests is becoming increasingly important. When we are surrounded by artificial things, it's harder to be peaceful. It's as if we begin to become artificial ourselves; we develop hypocrisy, suspicion and distrust. In that state, it's hard to develop genuine, warm-hearted friendships. We all feel the need to be surrounded by life. We need life around us that grows, flourishes and thrives. Because, as social animals, we also want to grow, flourish and thrive. We all love our technology. But our relationship with plants and nature is inextricably old and very deep. Buddhist ethics embrace all life, not solely human but also animal and plant life.

Franz Alt: The destruction of the environment has now also reached the Tibetan highlands. In the West, many people still dream of Tibet as a paradise, a Shangri-La. Is Tibet still a paradise?

Dalai Lama: What the Chinese did on the roof of the world after 1959 – especially during the Cultural Revolution – was cultural genocide. What I learn today from Tibetan refugees whom I meet in my exile in Dharamshala makes me fear that my home country is rather the opposite of a paradise. But I find it admirable that the vast majority of Tibetans, 70 years after the occupation, still cling to their religion, language and culture, and respect for their environment, even though there are more Chinese than Tibetans today living in Lhasa, Tibet's capital. The Chinese have reduced us to a minority in our own country.

Franz Alt: Do you think that you will reach your goal of returning to Tibet someday?

Dalai Lama: China is a great nation, an ancient nation – but its political system is a totalitarian system, not freedom. I am happy to live in India for the rest of my life. I can live in this country and utilise Indian freedom to fulfill my commitments towards the promotion of human values, religious harmony, the protection of Tibetan culture and environment, and the revival of ancient Indian knowledge.

Franz Alt: Should the Buddha return to our world and join a political party, he would certainly be Green, you say. What makes you so sure?

Dalai Lama: Buddha and we Buddhists have deep respect for nature and evolution. We know that nature does not need humans, but humans do need nature. Looking at today's global exploitation of nature, I think that without humans the Earth would do better [again laughing].

Franz Alt: Which political party would you support?

Dalai Lama: I have no hesitation in supporting initiatives that are related to protecting the environment. In Europe, I would vote for the Green Party, because threats to our environment are a question of our survival. This beautiful blue planet is our only home. It provides a habitat for unique and diverse communities. To take care of our planet is to look after our own home.

2. Ethics is more important than religion
Franz Alt: You would vote Green if you lived in a Western democracy. Why?

Dalai Lama: Because the Greens represent a similar nature-friendly philosophy as Buddhists do. For over a thousand years, nature has been sacred to we Tibetans. On the high Himalayan plateau where we live, we try, in the spirit of Buddhism, to live in peace with nature, protected by our mountains, without violence and in compassion with all living beings. Nature is sacred to us. Nature is our true home. We humans come from nature. We can live without religion, but not without nature. Therefore, I say that environmental ethics are more important than religion. If we keep destroying nature as we are doing today, we will not survive.

This is a law of nature that we have to accept. Humankind will suffer terribly if we do not learn that a clean environment is a human right, like any other human rights. It is our responsibility towards all sentient beings to ensure that we leave to our children and grandchildren a world at least as intact as we found it when we were born.

There are limits to what we are allowed to do, but no limits to our universal responsibility.

Franz Alt: What do you do yourself for the environment and for the climate?

Dalai Lama: On a personal and family level, we all need to develop a much clearer awareness of our actions and their consequences, such as how we use water or dispose of our rubbish, so that taking care of and limiting damage to the natural environment becomes part of our daily lifestyle. That is the proper way, and it can only be achieved through education.

I switch off the light when I leave my room. I take a shower instead of a bath. I eat little meat. I encourage other people to do the same.

As someone born in Tibet, the rooftop of the world, where Asia's great rivers start and world's highest peaks are to be found, I have loved nature since my childhood. I have made environmental conservation one of my life's commitments and advocate protection of the environment wherever I go.

For example, I honoured my promise to Sunderlal Bahuguna, the Indian environmentalist, to speak about environmental preservation. When I travel in the trans-Himalayan region from Ladakh to Arunachal Pradesh in India, I urge people there to plant trees to save their land from becoming barren in future. Trees create the landscape's greenery and bring peace and happiness of mind in our day-to-day life.

3. Vegetarianism helps the climate

Franz Alt: In 1965 you became a vegetarian. Have you been vegetarian since then and why?

Dalai Lama: In 1965, I became completely vegetarian: no eggs, nothing. Instead, I was eating lots of cream and nuts, and after 20 months, I had trouble with my gall bladder and got jaundice. My skin, eyes, nails – everything – turned yellow. My doctors advised me that I should go back to my original diet and eat some meat. I do so about once or twice a week. So, I am a little bit of a contradiction, telling people to be vegetarian as a non-vegetarian myself.

Nevertheless, right from the beginning, from the time I was in Tibet, I worked very hard to promote vegetarianism in Tibetan society. In the late 1940s, all the food served during Tibet's official festivals was vegetarian. Campaigns promoting vegetarianism have even been launched in the religious communities. In India, most of the Tibetan monastic institutions' kitchens have now started serving only vegetarian food to their monks and nuns.

Franz Alt: Worldwide meat consumption is rising, so more and more animals have to be killed. Do you think that worldwide meat consumption could drop again, and how is that to be achieved?

Dalai Lama: Buddhism does not forbid eating meat. It is a question of how and how much. Buddhism says no animals should be killed for eating. But our attitude towards meat is rather curious. Tibetan Buddhists can buy meat but should not kill animals.

What I find particularly worrying is intensive animal husbandry. We humans can largely live without or with little meat, and above all without animal suffering. Particularly in our modern world, where we have many alternatives, especially fruits and vegetables. In the meantime, there is even meat made from vegetables, for instance from peas and beetroot, from potatoes and coconuts.

Intensive animal husbandry has serious consequences, not only for animals but also for human health, the soil, insects and the air.

4. Buddhists disapprove of killing as a sport

Franz Alt: What you describe as 'curious' is also the attitude of most Western Christians towards meat. I am also only an 85 percent vegetarian. My doctor recommends eating meat or fish about once a week for health reasons.

If we had to slaughter animals ourselves for food, most of us would probably be strict vegetarians. That is also rather curious.

Meat consumption and animal husbandry cause about as many greenhouse gases globally as all cars, planes, trains and ships put together. The environmental physician Professor Hans-Peter Hüttner of Vienna Medical University also says: 'Meat plays a crucial role in the development of cancer of the intestine or circulatory diseases. Moderate meat consumption significantly reduces your disease risk, is beneficial to environment and climate and doesn't really hurt. A win–win situation.' What each person eats affects all of us.

What do you think about hunting and fishing as sports?

Dalai Lama: Buddhists disapprove of killing as a sport. I support those groups and people who work for animal rights and animal welfare around the world. It is sad that millions and millions of animals are killed for human consumption.

I once visited a poultry farm in Japan with 200,000 hens. They were kept in small cages just to produce eggs for two years. After that they were sold for slaughter. That was shocking. We should support people who fight against such unworthy businesses and such animal misery. It is also very dangerous and short-sighted for us simply to suppress and forget about animal suffering. What we are doing to animals today can also happen to us. Perhaps one day we will kneel down and ask the animals for forgiveness. I also disapprove of the way in which we have mechanised farming these days.

We must never forget the suffering we inflict on other sentient beings. I am thinking of some Tibetan butchers or Japanese fishermen who ask the animals they kill for forgiveness and pray for them.

Franz Alt: Do you think that worldwide meat consumption could drop again, and how is that to be achieved?

Dalai Lama: In some countries this is already the case. I meet young people everywhere who are looking for alternatives to brutal meat consumption. In the United States, the explosion of new meatless products produced and distributed by companies such as 'Beyond Meat' has grown into a wider movement encouraging everyone to eat more plant-based diets. Many consumers want to reduce their meat consumption in order to protect the climate, but also to alleviate animal suffering caused by factory farming. Now there are vegetarian 'hamburgers' [laughing].

Franz Alt: But that means that we Westerners would have to learn to let go at least a little bit? Is letting go the heart of greater environmental justice?

Dalai Lama: Yes, you could say so. Letting go of surplus is the heart of spiritual growth. Just imagine what we could achieve if the United States halved its military budget. That would free up $300 billion every year for environmental projects such as the solar-energy transition or overcoming hunger in poor countries. Defending the future instead of dangerous military upgrading. That could indeed be the beginning and impetus for an ecological age. 'Letting go' would mean liberation.

Franz Alt: Russian novelist Leo Tolstoy said: 'As long as there are slaughterhouses, there will be battlefields.' Do you also see a connection between intensive livestock farming and violence among people?

Dalai Lama: Such a connection exists. In all religions we know this spiritual law: 'As you sow, so shall you reap.' By nature we have inner inhibitions about killing. Above all, we feel that it is not right to inflict pain on other sentient beings. If we let our conscience be brutalised while killing animals, it will also be brutalised while killing people.

Franz Alt: 'Whoever changes, changes the world', you say. Western societies are mainly concerned with today's generations: past generations no longer exist, and future ones are not yet imagined. Given that only a minority in the West believe in rebirth, what chance do environmental and climate protection have?

Dalai Lama: As I already said: the belief in reincarnation can help to protect the environment. We must not leave young people alone in their fight for a good environment and climate. It is important that we organise and publicise and have global solidarity with young people. Everybody and everything must change if we want to live in a climate that is compatible with life. Our generation has damaged the climate, so we also should help to save it.

Franz Alt: For many months now, hundreds of thousands of young people have been demonstrating for better climate protection in more than 100 countries. Do these young demonstrators give you hope?

Dalai Lama: Members of the younger generation to whom the 21st century belongs have important responsibilities: they must learn from and rectify the mistakes of the past, and ensure that such mistakes are not repeated. The younger generation, who will inherit this Earth, has the ability and the opportunity to act and create a more compassionate world.

Franz Alt: Your Holiness, dear friend. I cordially thank you for these reflections, which we have been exchanging for decades. They will help many people to understand that our 21st century must become the one in which global humankind finds ways to accept universal responsibility. I have learned from you that every single one of us can and must take on their own piece of universal responsibility if we want a better world. Inner peace, love and compassion are the most important energies that will lead to external peace, too, and to peace with nature.

Whenever the Dalai Lama and I say goodbye, he places around my neck a kata, a greeting scarf made of white silk with the traditional sign of good luck. He takes my head between his hands, our foreheads and noses touch one another in friendship and we embrace for a long time and sense that love and peace between us humans is possible. It is the spirit that gives strength to peace, justice and friendship.

7

The Sheltering Tree of Interdependence – A Buddhist Monk's Reflections on Ecological Responsibility

This poem was released on the occasion of the presentation by His Holiness the Dalai Lama of a statue of the Buddha to the people of India and to mark the opening of the International Conference on Ecological Responsibility: A Dialogue with Buddhism on 2 October 1993 at New Delhi. (A booklet of the poem, in Tibetan and English, is distributed by Tibet House, New Delhi.)

'During the course of my extensive travels to countries across the world – rich and poor, East and West – I have seen people revelling in pleasure and people suffering. The advancement of science and technology seems to have achieved little more than linear, numerical improvement: development often means only more mansions in more cities. As a result, the ecological balance – the very basis of our life on Earth – has been greatly affected.

'On the other hand, in days gone by, the people of Tibet lived a happy life, untroubled by pollution, in natural conditions. Today, all over the world, including in Tibet, ecological degradation is fast overtaking us. I am wholly convinced that, if all of us do not make a concerted effort, with a sense of universal responsibility, we will see the gradual breakdown of the fragile ecosystems that support us, resulting in an irreversible and irrevocable degradation of our planet, Earth.

These stanzas have been composed to underline my deep concern, and to call upon all concerned people to make continuous efforts to reverse and remedy the degradation of our environment.

1. O Lord Tathagata
   Born of the lksvakus tree
   Peerless One
   Who, seeing the all-pervasive nature

Of interdependence
Between the Environment and sentient beings
Samsara and Nirvana
Moving and unmoving
Teaches the world out of compassion
Bestow thy benevolence on us

2. O the Savior
The one called Avalokitesvara
Personifying the body of compassion
Of all Buddhas
We beseech thee to make our spirits ripen
And fructify to observe reality
Bereft of illusion

3. Our obdurate egocentricity
Ingrained in our minds
Since beginningless time
Contaminates, defiles and pollutes
The environment
Created by the common karma
Of all sentient beings

4. Lakes and ponds have lost
Their clarity, their coolness
The atmosphere is poisoned

Nature's celestial canopy in the fiery firmament
Has burst asunder
And sentient beings suffer diseases
Unknown before

5. Perennial Snow Mountains resplendent in their glory
Bow down and melt into water
The majestic oceans lose their ageless equilibrium
And inundate islands

6. The dangers of fire, water and wind are limitless
Sweltering heat dries up our lush forests
Lashing our world with unprecedented storms
And the oceans surrender their salt to the elements

7. Though people lack not wealth
They cannot afford to breathe clean air
Rains and streams cleanse not
But remain inert and powerless liquids

8. Human beings
And countless beings
That inhabit water and land
Reel under the yoke of physical pain
Caused by malevolent diseases
Their minds are dulled

With sloth, stupor and ignorance
The joys of the body and spirit
Are far, far away

9. We needlessly pollute
The fair bosom of our Mother Earth
Rip out her trees to feed our shortsighted greed
Turning our fertile earth into a sterile desert

10. The interdependent nature
Of the external environment
And people's inward nature
Described in tantras
Works on medicine, and astronomy
Has verily been vindicated
By our present experience

11. The earth is home to living beings;
Equal and impartial to the moving and unmoving
Thus spoke the Buddha in truthful voice
With the great earth for witness

12. As a noble being recognises the kindness
Of a sentient mother
And makes recompense for it
So the earth, the universal mother

Which nurtures equally
Should be regarded with affection and care

13. Forsake wastage
Pollute not the clean, clear nature
Of the four elements
And destroy the well-being of people
But absorb yourself in actions
That are beneficial to all

14. Under a tree was the great Saga Buddha born
Under a tree, he overcame passion
And attained enlightenment
Under two trees did he pass in Nirvana
Verily, the Buddha held the tree in great esteem

15. Here, where Manjushri's emanation
Lama Tsongkhopa's body bloomed forth
Is marked by a sandal tree
Bearing a hundred thousand images of the
    Buddha

16. Is it not well known
That some transcendental deities
Eminent local deities and spirits
Make their abode in trees?

17. Flourishing trees clean the wind
    Help us breathe the sustaining air of life
    They please the eye and sooth the mind
    Their shade makes a welcome resting place

18. In Vinaya, the Buddha taught monks
    To care for tender trees
    From this, we learn the virtue
    Of planting, of nurturing trees

19. The Buddha forbade monks to cut
    Cause others to cut living plants
    Destroy seeds or defile the fresh green grass
    Should this not inspire us
    To love and protect our environment?

20. They say, in the celestial realms
    The trees emanate
    The Buddha's blessings
    And echo the sound
    Of basic Buddhist doctrines
    Like impermanence

21. It is tree that brings rain
    Trees that hold the essence of the soil
    Kalpa-Taru, the tree of wish fulfillment

Virtually resides on earth
To serve all purposes

22. In times of yore
Our forebears ate the fruits of trees
Wore their leaves
Discovered fire by the attrition of wood
Took refuge amidst the foliage of trees
When they encountered danger

23. Even in this age of science
Of technology
Trees provide us shelter
The chairs we sit in
The beds we lie on
When the heart is ablaze
With the fire of anger
Fueled by wrangling
Trees bring refreshing, welcome coolness

24. In the trees lie the roars
Of all life on earth
When it vanishes
The land exemplified by the name
Of the Jambu tree
Will remain no more than a dreary, desolate desert

25. Nothing is dearer to the living than life
    Recognising this, in the Vinaya rules
    The Buddha lays down prohibitions
    Like the use of water with living creatures

26. In the remoteness of the Himalayas
    In the days of yore, the land of Tibet
    Observed a ban on hunting, on fishing
    And, during designated periods, even construction
    These traditions are noble
    For they preserve and cherish
    The lives of humble, helpless, defenceless
        creatures

27. Playing with the lives of other beings
    Without sensitivity or hesitation
    As in the act of hunting or fishing for sport
    Is an act of heedless, needless violence
    A violation of the solemn rights
    Of all living beings

28. Being attentive to the nature
    Of interdependence of all creatures
    Both animate and inanimate
    One should never slacken in one's efforts
    To preserve and conserve nature's energy

29. On a certain day, month and year
    One should observe the ceremony of tree
        planting
    Thus, one fulfills one's responsibilities
    Serves one's fellow beings
    Which not only brings one happiness
    But benefits all

30. May the force of observing that which is right
    And abstinence from wrong practices and
        evil deeds
    Nourish and augment the prosperity of the
        world,
    May it invigorate living beings and help them
        blossom
    May sylvan joy and pristine happiness
    Ever increase, ever spread and encompass all that is

8

For a Solar Age – Epilogue
by Franz Alt

1. Reconciling economy and ecology

Climate change is not a distant threat: it's already here. We have to face reality, which might be complicated but is not hopeless. Since the Enlightenment about 300 years ago and the 'de-idealisation' of the world, our Western cultural model has been based on scientific knowledge – or so we like to think.

Why have scientists worldwide been warning for decades against climate change without actually being heard in politics and society, let alone bringing about appropriate actions? Why is the Enlightenment of the past not sufficient to secure our salvation? In order to prevent the worst effects of climate change, we need a second Enlightenment, more profound … an 'enlightening' of the Enlightenment.

In this book, the Dalai Lama demonstrates very clearly that today's environmental crisis is the crisis of our inner world.

We think we know what we are doing. In fact, we are not doing what we know. The idea that rationalism alone will save us is irrational. Reason alone will not bring humanity to its senses. We love to suppress such insight. There are those, however, who are fully aware of what they are doing.

Many are afraid of the necessary changes. It is true that sometimes politicians do adopt necessary changes because of fear, such as banning CFCs because they can cause skin cancer or introducing the three-way catalytic converter in German cars because of a fear of dying forests.

The lobbyists of the traditional energy industry are simply afraid of losing their benefits. Politicians meanwhile rely on the greed of big companies, while many individuals continue to travel by plane or car and eat themselves sick with a lot of meat, even though they know what they are doing to themselves, the environment and their children's future. Can humans really confront this reality? Is the classical Enlightenment – the 'Age of Reason' –sufficient for us?

Spiritual and religious people have trusted the wisdom of nature for millennia. But this confidence is being shaken in times of global warming, pandemics and the extinction of species. Mother Nature no longer cooperates but instead runs riot; she has got a fever and is on strike. We are about to lose humanity's partner and confidante – nature, the source of our wealth and happiness. But first we should at least understand that we do not want to save the climate, as such, but – quite egotistically – ourselves.

The second Enlightenment will succeed only by combining religion and philosophy, nature and reason, freedom and responsibility. This is what we call eco-spirituality. In this book, the Dalai Lama also speaks of education of the heart. If we do not understand this interdependence, our liberty will soon end in enslavement. Global warming is already creating a lack of freedom, for instance for refugees or farmers who are losing their land to the desert or for elderly Europeans who died during the hot summers of 2003 and 2018. In 2003 alone, about 60,000 people died of heat within the European Union. In India, where temperatures rose to an unbearable 50 °C (122 °F), hundreds of thousands of elderly people died from overheating. It is bewildering to see how difficult it is to understand these correlations, especially for conservatives. Helping to preserve the creation should be conservatives' central task.

I wish the conservatives were really conservative!

After the Enlightenment, many intellectuals believed that humans could be liberated not only from our self-inflicted immaturity but also from nature. Within only 300 years, we have robbed nature of the resources she gathered during 300 million years, blew the waste into the air and filled the gigantic holes that were made with waste. Today's economists call this progress. What now?

2. There is no matter

For 300 years, economists have believed that money is the basis of all economic systems. Nature, however, is the basis of any economy truly worthy of the name. An economy based on dead soil makes little sense and does not need any jobs.

The words *economy* and *ecology* both derive from the Greek *oikos*, meaning housekeeping or household management. The power of money is the principal disease of our time. 'The attempt to separate numbers from values results in total dominance of the numbers', writes Christian Felber, Austrian economist and author, who coined the term 'Economy for the Common Good'. And: 'Separating economy from ecology is one of the greatest sins committed by economists.'

The increase of money does not make the world richer. In the past 50 years we have brought about the most brutal kind of poverty by killing off about half Earth's species of animal and plant life. More money, but less richness of life: compared to this crazy economics, theology is almost an exact science. Today's economists should learn to place their science in a broader, holistic context. Pope Francis says: 'Everything is connected with everything: the poor and nature are crying for help.'

Joachim Schellnhuber, climatologist and physicist, has reached a similar conclusion: 'Speaking to an economist is the worst punishment for a physicist.'

There will be no future unless we learn now to avoid the biggest mistakes of the past. Only then will we arrive in the real world. Or in the divine order.

In religious language, God is identical with spirit, according to St. John's Gospel. With this in mind, Nobel laureate Max Planck said:

'As a physicist, that is, as somebody who has devoted his whole life to the most clear-headed science, to the study of matter, I will certainly not be held to be a dreamer. I can tell you as a result of my research about atoms this much: there is no matter as such. All matter originates and exists only by virtue of a force which brings the particle of an atom to vibration and holds this most minute solar system of the atom together. We must assume that behind this force is the existence of a conscious and intelligent Mind. Not the visible, yet transient matter is real and true, but rather the invisible and immortal spirit. As there cannot be a mind as such and as every mind belongs to a being, we are forced to assume the existence of spiritual beings. As these beings cannot exist by themselves but must be created, I have no hesitation in calling this mysterious Creator, like all ancient civilisations, God!'

Professor Hans-Peter Dürr, physicist and former Director at the Max Planck Institute in Munich, argues similarly in his popular book *There Is No Matter*. Shortly before he died, we walked up to the Acropolis in Athens. He already had great difficulty breathing. Surrounded by so many stones and rocks, I said to the world-renowned physicist: 'Everything here is actually matter.' 'Oh no,' he replied (to the dismay of many a Marxist and materialist), 'All this is materialised spirit. The spirit has been, and will always be, primary.' A surprising insight for a physicist.

3. In depth all life is one

Hans-Peter Dürr's research shows remarkable parallels between Judaeo-Christian thinking, Hindu-Buddhist insights and the latest findings of modern quantum physics. We must finally learn to cross borders in order to reconcile what seems irreconcilable. The borders of pure rationalism and the Enlightenment exist at the surface: in depth, all life is one.

This is the conclusion of both contemporary quantum physicists and the Dalai Lama, according to the German physicist Carl Friedrich von Weizsäcker.

If we understand, or imagine, God as the Sun behind the Sun, solar thinking and acting will be more than a purely technical transition from the fossil-fuel age to a solar age; the transition will also be a sign of a new, profound, holistic and mature attitude towards life of divine substance.

Tens of thousands of people worldwide have been inspired by this attitude and change of mind. They say to me: 'We have now taken a new attitude towards nature und the Sun. We're looking upwards more often and also understand what Jesus meant in the Sermon on the Mount, when he said, "He causes his Sun to rise on the evil and the good."'

No Sun exists solely for energy companies. The Sun is a divine symbol that shines on all of us, a master craftsman creating inexhaustible master-pieces and inspiring the wonders of the mind.

4. No child should starve to death

To many 'solar humans', as I term those whose eyes have turned to the future, looking upwards also means looking inside. They learn to gain access to their souls through their dreams. Thanks to their new thinking about external energy, they become aware of their inner energy. They recognise that the external energy crisis corresponds to a much deeper internal energy crisis, a crisis of their soul. And the human soul, in the words of the renowned Swiss psychologist Carl Gustav Jung, 'is the only super-power of this world that I recognise'.

If we acted in a way that was truly holistic and based on knowledge, with an awareness of our soul, we would have reduced the use of greenhouse gases a long time ago. We would not only have spoken about but actually realised the transition in the field of energy and transport, water and agriculture. But we are living in a period that, to a great extent, is unaware of our souls, which makes us resistant to, even incapable of, learning. It seems that only inner healing can bring about outer healing. As the mystics of the Middle Ages already knew, outer like inner, inner like outer.

What Jesus called our 'inner light' can help us become a sunny being on the outside. And we may suddenly sense that the most important things in life are free: joy and gratitude in ourselves, as well as in the Sun above us. Yet joy, love and gratitude are no more commodities than the Sun is. They are God's gifts. Let us finally make intelligent use of them. For this purpose we should open our hearts as landing strips for the spirit from above, in the same way as we make the roofs and walls of our homes available for solar installations. So we can – perhaps for the first time in humankind's history – work together so that one day children will no longer have to starve and people will not be forced to flee their countries. And we will learn that nobody leaves his or her home voluntarily. Energy crisis, refugee crisis and climate crisis are closely linked. If we see this correlation, we will find solutions: crises always give rise to new opportunities. The key to the solution of all these crises is the energy crisis.

In October 2018, I was invited to give lectures in Mali, West Africa. In Bamako, Mali's capital, I was asked to speak at an African solar energy conference. Africa and the Sun. What an opportunity! At the invitation of the Minister of Energy, we speakers travelled to a village of 20,000 inhabitants, outside the capital. Only three years ago, there had been no electricity, but they now had solar installations and electricity. The village pharmacist told me that solar energy had raised the health level of the village. Unlike before, he could now keep his medicines cool. In a school, I met children who enthused about the benefits of solar energy, because now they could watch soccer on TV. Mothers reported that they could now send their children to school because the children could do their homework in the evening, by the light of solar lamps, which was impossible before. Education changes everything. A dressmaker told me that she at last had an electric sewing machine and no longer had to slave away at the treadle. The village mayor said to me: 'Since we've had electricity in the village, young people don't think of fleeing to Europe any more. Solar energy also creates new jobs.'

All problems created by humans can also be solved by humans.

Shortly afterwards, I gave a lecture at the Conference of the World Wind Energy Association in Karachi in Pakistan. Here the Chinese company Goldwind, one of the world's largest producers of wind turbines, has — with the help of German technology — set up a wind park that produces affordable and clean electricity for 1.5 million people in the city. My Pakistani friends are full of praise for such progress. They are fully convinced that their country will achieve the complete energy transition by the middle of this century.

Karachi and Mali could be everywhere.

Every wind turbine and every solar installation, every hydropower plant and every biogas plant is a sign of peace. Never are wars waged over the Sun or the wind.

5. Disarming instead of rearming

Can this trend be financed? This is a typical and immediate objection in European countries. Today, we spend about $1.6 trillion a year worldwide on weapons and the military, in accordance with the principle of ancient Rome: 'If you want peace, you must prepare for war.'

The result of this traditional way of thinking? Two thousand years of war, misery, destruction and annihilation. Millions of casualties. In the atomic age, it is high time we updated the motto: 'If you want peace, you must prepare for peace.' This means disarming instead of rearming. Solar energy transition can be financed worldwide with just a small fraction of the money spent on war preparations. We are still caught in the old war trap.

So what are we waiting for? We humans can learn and change our ways. As early as 1972, the Club of Rome commissioned a much-discussed report, *The Limits to Growth*. At the time, we produced a special broadcast on the topic in *Report Baden-Baden* in Germany, and reached an audience of millions. For quite some time, the newspapers covered this crucial topic in detail. The book became a bestseller and was translated into numerous languages.

We have known for a long time that there are limits to physical growth. Humans, for example, grow until we are 160, 170, 180 or 190 centimetres tall, or perhaps a little taller. But that is enough. Nobody grows forever. After physical growth ends, something else matters: inner growth and inner maturity.

We cannot carry on growing physically, but as far as spirit, mind, culture or religion are concerned, we can continue to mature throughout our lives. Our material resources are limited, unlike the ideas of our mind. Ideas can multiply as if they had sexually reproduced. This is the basis of progress and prosperity. Nevertheless, governments worldwide continue to pursue endless economic growth – yet the only thing that grows without limits is cancer. Our philosophy of eternal growth is propagating a quasi-official cancerous economy, and is doing so worldwide. This could prove fatal.

But how can we, in times of global warming and environmental destruction, grow spiritually so that we will be up to these challenges before it is too late? How can we mature rather than simply grow? This may be the question of all current questions.

The greatest lie in politics: 'But we are already doing so much.' In my TV series *Zeitsprung*, we showed that, in 1993, humankind was discharging about 20 billion tonnes of greenhouse gases into the atmosphere every year. And as if this were not enough, it is today almost 36 billion tonnes per year – after more than 20 world climate conferences with tens of thousands of participants, more than 27 years later.

But we are already doing so much, aren't we? Yes we are – but it's all wrong!

Every day we are emitting globally about 136 million tonnes of CO_2. Every single day we kill off about 150 animal and plant species, we lose 45,000 tonnes of fertile soil and deserts grow by about 80,000 hectares. Our greed for meat destroys rain forests. In 2019, hundreds of millions of Africans lived through the most terrible drought in living memory and are fearful of the next famine. A whole subcontinent is crying for water: Angola, Botswana, Congo, Lesotho, Malawi, Mozambique, Namibia, Rwanda, Zambia, Zimbabwe, South Africa. Is there still hope for us?

6. Managing with nature, not against it

We are experiencing not ecological standstill but a speedy step backwards. We are racing towards the abyss, with politicians cheering us on: 'Speed up! It's the only way to save us. Growth, growth, growth.' This is perverse in the extreme. But how can we turn away before everything keels over and we finally topple into the abyss?

We are destroying our Earth because we are insanely concerned with the present and the future. The motto of our Western cultural model remains that greater always means better. The fact that more and more ever-larger SUVs are being bought in Germany is a case in point. In our cities, an increasing number of these vehicles weighing 2.2 tonnes and measuring 470 by 180 centimetres (5.2 x 2 yards) are running against the world climate, although space in the cities is getting tighter and tighter. Owners and drivers – men and women alike – are more concerned with their egos than with actually getting around.

Most people couldn't care less about the future. Having just concluded a speech on energy transition, I once had this experience: an elderly gentleman came to the book table, saying: 'Well, Mister Alt, you may be right about solar and wind energy, but, you know: I'm seventy-five now, so there's enough for me.' 'Have you got children?', I retorted. He bowed his head, dumbfounded, and went away.

A widespread approach seems to be: 'After us, the deluge'. We may be the first generation who can no longer say to their children in all conscience: 'We love you.' Many children would have to answer their parents: 'We don't believe you. This is hypocritical. You're only pretending. If you really loved us, you wouldn't burn our future.'

Many children and young people have begun to see through us. They are revolting against our pyromania. Immediate solar energy transition has become a matter of survival for humankind. It is not just the young. The Fridays for Future movement is supported by as many as 28,000 climatologists from Scientists for Future, as well as by Parents for Future, Grandparents for Future, Farmers for Future, Doctors for Future and Entrepreneurs for Future, not forgetting the first Journalists for Future, Churches for Future and even Climbers for Future.

This is entirely consistent with the Dalai Lama's position and absolutely in the spirit of the Pope's Encyclical '*Laudato si*'. What the world needs now is a movement called Citizens for Future. The crucial question is: How can real change be brought about? The answer is: Not through fear but with positive emotions such as compassion and mindfulness.

Without energy, economic development is impossible in countries that are still poor. Without energy, people in emerging countries have no option but to flee to economically rich countries. What would we do if we lived in poorer countries and saw no prospects for our children?

According to a UN prognosis, there will be more than 400 million climate refugees by the end of the current century. They will be fleeing to countries where they see economic prospects. We will reap what we have sown. We, the industrialised countries that have caused climate change. The poor are the victims of our actions. And that is why they will come here, unless we stop global warming. A Bangladeshi or a sub-Saharan African consumes about 20 times less energy than a German. We are responsible for this situation, not the Africans who want to, or have to, come to Europe. Where else should they go?

On average, a US citizen emits 16 tonnes of carbon dioxide per year, a German 8 tonnes, a Swede 4 and a Bangladeshi or sub-Saharan African 0.4 tonnes.

So far, it has been mainly war refugees that have come to Europe, such as those from Syria or Afghanistan. Usually, they return after the wars to rebuild their home countries, like the refugees from the former Yugoslavia after the wars in the 1990s. But where should future climate refugees return to?

Global warming concerns us all. It is a problem that we are again about to bury once the pandemic has run its course. But everything we repress will one day come home to roost. Millions of climate refugees will make their way to Europe in the future unless we put an end to the cause of their flight, global warming. Seriously and immediately.

Climate change is, as I've already said, a war against nature that concerns all countries. We will have to learn to distinguish between war refugees and those forced to flee by climate change. Theirs is a flight of no return.

The history of humankind is a history of refugees. In a way we are all refugees. This began about 200,000 years ago, when *Homo sapiens* left Eastern Africa to conquer the whole world.

I am writing these lines on board Germany's ICE, a high-speed train, on 14 September 2019, the 250th anniversary of the birth of the German naturalist Alexander von Humboldt. Travelling in Latin America from 1799 to 1804 for the purpose of research and discovery, Humboldt gained a myriad of new insights into the laws and richness of nature.

Never before had reports on natural history attracted so much worldwide attention. Today, on his anniversary, many German newspapers are calling Alexander von Humboldt the 'world's first environmentalist'. Humboldt enthused about the 'wonders of the lush primeval forests' and their biodiversity. He was 'out of his senses', he wrote excitedly. This brilliant scientist, always eager to learn, would today be 'out of his senses' again, were he to witness the unbelievable and disgusting brutality that is destroying his 'wonders of nature' – out of sheer greed and ignorance.

As a universalist and naturalist, Humboldt saw the world as a whole. Nowadays, when the findings of scientists who think and do research on a holistic basis are ignored or denied, Humboldt could serve as a model. What he considered as holistic is what the Dalai Lama in this book calls interdependence.

Humboldt's enthusiasm for flowers and leaves, for the rivers and flies of the tropical rain forests, together with his scientific meticulousness, is contrary to where the spirit of maximal exploitation and brutal greed has led us. This evil zeitgeist has set Amazonia ablaze. In autumn 2019, so much forest in Sumatra was burning that entire cities almost suffocated, thousands of schools had to close and the blaze released 326 million tonnes of carbon dioxide in five weeks. The glacial ice is melting in Greenland as well as in Alaska, in the Arctic and Antarctic, in the Alps and in the Himalayas. Permafrost in Siberia is thawing. In the past few decades we have destroyed half of the rain forests, the lungs of our planet. Only one of its lobes has survived.

Together with the primeval forests, the idea of animated nature, in which everything is interrelated, also dies. Thus we are destroying the basis of our own life. US president Trump and the German political party AfD (Alternative for Germany) as a whole deny global warming, contrary to the views of all experts.

Alexander von Humboldt never distinguished between knowing and sensing, between feeling and reason, between humans and nature, between economy and ecology. He knew that humans are dependent on the climate. Humboldt's cross-linked knowledge was always coupled with a sense of responsibility for the world as a whole. As historian Kia Vahland wrote in the *Süddeutsche Zeitung*, he embodies an 'energetic universalism, abounding in knowledge', an exemplary model for our time.

Tackling global warming and the threatening worldwide climate catastrophe is the most vital and significant challenge for humans this century and probably far beyond. Global warming affects everybody in every country. As with the Covid-19 pandemic, we all have a common enemy. This challenge could, and should, unite us in the fight against the greatest of our enemies, global warming.

It is no overstatement but a simple truth: the end of our civilisation has become possible. And therefore, on the occasion of the world strike day on 20 September 2019, 200 international media organisations united for the first time to inform people about this global threat. So nobody can pretend any longer that he or she didn't know.

On 20 September 2019, the Dalai Lama wrote: 'Young people worldwide are demonstrating for a good climate, which is wonderful. Thereby they demonstrate a realistic view of their own future. We adults should support these young people.' Greta Thunberg's opponents accuse her of sentimentality and irrationality, although science and reason are on her side, not that of her detractors. Naturally, global warming is a highly emotional issue. During her speech at the UN, Greta unleashed tears, anger and despair. But Greta's opponents argue far less rationally.

7. Deeds are evidence of the truth

Transformation is possible: there are always alternatives, as we have already seen. We can do something. We do not have to leave things as they are.

'There's nothing I can do about it.' This is the most common and fatalistic excuse when people can't think of anything better to say. It is nothing more than an excuse for inaction.

Everybody is by nature capable of transformation. This is the reason why we are here. All problems caused by humans can also be solved by humans.

+ Peace is possible
+ Love is possible
+ Justice is possible
+ Compassion is possible
+ Climate protection is possible
+ Sustainable economy is possible
+ A better world is possible

Fine words must of course be followed by appropriate deeds. Deeds are the evidence of our truthfulness. Only then will utopias become concrete, realised visions.

Greed for money is born of ignorance, according to both Buddha and Jesus. Greed, it follows, is irrational. No amount of money, no share price, no GNP, no property can ever be enough to satisfy our greed and end our irrational pursuit. People living in the rich industrialised countries, whose income has more or less doubled every 20 years since 1945, are not happier than their forebears before 1945.

The only antidote to money and greed is compassion, as taught by Buddhism and early Christianity (for as long as it followed the teachings of Jesus rather than those of the Church Fathers). At their roots, both religions ignore dogmatism; they are committed to pragmatism and science. If science invalidates the scriptures, the scriptures have to be rewritten, even the so-called Holy Writ.

Masterpieces of literature and, for that matter, fairy tales are imbued with the idea of the human ability to change: *The Divine Comedy* by Dante, Goethe's *Faust*, *The Odyssey*, the Gilgamesh epic,

Parsifal, Jesus's 'Sermon on the Mount', Plato's 'Cave Parable', or Mozart's *Magic Flute*.

Thanks to 20th-century psychology, as pioneered by Sigmund Freud and Carl Gustav Jung, as well as neuroscience of the early 21st century – neuropsychology, neurophilosophy and neurobiology – we have recognised that humans in principle are able to change and to alter. Jung calls such processes of alteration 'individuation' or 'self-actualisation'. For Jung, individuation means 'anima-integration' to the man who tries to integrate the female parts of his soul, and 'animus-integration' to the woman integrating the male parts of her soul. In Jung's view, the 'self' or 'individuation' stands for 'unity and wholeness of the complete personality'.

Alteration – or, in religious terms, 'conversion' – is always possible. People can learn, but only if they want to. Our will may often be blind, but it is not stupid. We can train it like a muscle. This is why, over the course of history, what once seemed impossible has often become possible: the abolition of slavery and child labour, women's emancipation and the separation of state and church, human rights and democracy and, in 1989, German reunification.

9

Ten Commandments for the Climate

+ By 2035 at the latest, greenhouse gases must be reduced to zero. The most effective way to protect the climate is the immediate abandonment of coal-fired power. Slovakia plans to phase out coal by 2030, Greece by 2028 and England, a traditional 'coal country', as early as 2025. Why Germany, the US and the UK not before 2038?
+ All new buildings must be emission-free, for instance, by using more timber constructions. As a building material, aluminium damages the climate 128 times more than wood. More and more Europeans are using wood for their houses.
+ From now on, the construction of power plants should only be authorised if

renewable energies are used. Cut today's billion-dollar subsidies for industrial polluters.

✦ As of 2025, only electric cars or other vehicles with CO_2-free engines should be licensed. Such a measure works, as California showed in the 1990s by introducing quotas on electric vehicles. China, the world's biggest automotive market, plans to do so. Now all the others must follow.

✦ Public transport must be expanded significantly. More conferences must take place via Skype or Zoom instead of in face-to-face meetings, a development already seen in earnest during the pandemic. Houses, streets and industry should occupy less space; we need to make our cities more dense by building upwards. Building ecologically does not mean new buildings, but primarily redeveloping and renovating. New industrial plants should be CO_2 emission-free from 2025 onwards. A deadline for the compulsory introduction of zero-emission technologies will drive the necessary innovations worldwide.

✦ About 25 percent of annual greenhouse gas
  emissions are caused by the production of
  food, especially meat products. Producing
  beef soup causes 10 times more greenhouse
  gases than producing vegetable soup. Does
  meat soup really taste 10 times better than
  vegetable soup? All of us should consider
  and follow dietary guidelines, such as those
  outlined in the Eatwell Guide released by the
  NHS, which has begun to introduce elements
  of sustainability into its guidance and
  suggests we reduce our meat consumption.
  Doing so helps to prevent obesity and high-
  blood pressure, slows down climate change
  and reduces nitrogen levels in groundwater.

Climate change must also be understood as a medical emergency. The correlation between climate
change and human health has so far received far
too little attention. Global warming is the greatest
threat to health in the 21st century, according to the
World Medical Association. And emissions from the
burning of fossil fuels increase the risk of a stroke,
asthma and diabetes. Global warming is fatal.
Conversely, we can choose to act to protect the

climate while also improving our health. By riding a bicycle or walking – another change hastened by the pandemic – we not only take good care of the environment but also reduce the risk of cardiovascular diseases, diabetes and being overweight. Burning less coal means less fine dust and less lung disease.

Unlike other religions, Buddhism does not grant humans a higher right to life than other living beings. A Buddhist monk would never say, as did the medieval Christian monk Saint Thomas Aquinas, 'Animals have no soul'. Neither would Jesus have used those words. He advocated compassion for all creatures. In the New Testament, I found 16 animal species mentioned in his parables.

✦ We must reforest worldwide and green the deserts, as the child and youth organisation Plant for the Planet has been doing for years. They have already planted 14 billion trees, with the aim of 1,000 billion trees. Young trees will not help immediately, but at least something is growing. Pakistan has announced that it will plant 10 billion trees by 2030. Ethiopia, even poorer than Pakistan, holds the world record: in the summer of 2019, more than 350 million young trees were planted on a single day!

+ We must vote exclusively for politicians who honestly represent our interests rather than those of obsolete fossil–nuclear energy or of the fossil automobile industry. Democracy instead of autocracy, Sun instead of atom and coal.

+ Solar development in poor countries is the best precaution against population growth that is out of control.

+ We can all make fewer purchases and reduce waste, ride a bike more often or go jogging, be more environmentally friendly whenever we have a party, switch to green electricity, invest our money in an ecologically conscious and green way.

Ultimately we should do what we deem right – what we know to be right. We must live in a simpler way so that others will simply survive. We must think more and offer resistance against ignorance and short-sightedness. We must free ourselves from overabundance.

10

What Can I Do?

1. Choose wisely

It takes great effort to realise these Commandments. But in the end it will mean a better life for all of us, a life worth living. The fruit of climate justice is peace. In order to overcome an obsolete, materialistic worldview, we need a positive vision that is more attractive. The world revolution for compassion, as suggested here by the Dalai Lama, can be of great, perhaps even decisive, help. The time seems ripe. Large parts of the younger generation, and increasingly of older ones, seem prepared.

Who says that an individual can't do anything about it? If each and every person puts his or her own house in order, the whole world will be a better place. 'The future depends on what you do today', said Mahatma Gandhi. Who could stop us, apart from ourselves? A better world begins with each individual.

As a species, we are in danger of disappearance unless we learn that a healthy forest ensures our own health. Perhaps we humans need to develop a greater awareness of trees. Novelist Richard Powers says we should immediately abandon our blind belief in the supposed 'special status of a human being'. An 'awareness of plants' could be helpful. These terms are pretty close to the Dalai Lama's Buddhist thinking and his holistic message about the 'Revolution of Compassion' as well as Albert Schweitzer's 'Reverence for all Life'. Whoever remains alert cannot deny the threatening catastrophe any longer. How far are we prepared to go in order to stop the apocalypse? Whoever is responsible for a problem can also put things right – this is probably the most important thesis of this book. As a human family we share a common destiny. So let's take care that it's not getting too hot.

We decide in favour of, or against, sustainable building; we decide whether we travel 'green' or in a way that damages the climate; whether we live on food that destroys resources or on the produce of organic farming; and whether we use renewable or fossil-atomic energy. The transformation we need for a good future for everyone has already taken place among millions of people who serve as our models. This transformation is no longer a dream; it is often becoming real.

Rolf Disch, a solar architect, has been building houses for 25 years that produce three times as much solar electricity as the occupants need, as well as enough energy for heating. His house earns money. This is the solar future. Disch's own house produces more than six times as much solar electricity as he and his wife consume. I know of social housing powered by solar energy. The occupants now spend half as much on energy as they did in the past on traditional energy. For such projects, solar electricity is social electricity.

This book aims to encourage action: political action as well as private, personal action, including voting habits, in favour of the environment and climate, which is the foundation of our lives. Our voting behaviour reflects our political responsibility in a democracy.

Earlier in this book, the Dalai Lama says that in Europe he would vote for the Green Party. I am also convinced that, without the Greens participating in the federal government, Germany will miss its climate protection targets in 2030, as was the case in 2020.

I say this as someone who was a member of the Christian Democrats for 28 years. The traditional parties are still stuck in obsolete ideas, thinking of the old energy and automobile companies.

All technical requirements needed to overcome the fossil-atomic age have been met. Our problem is not a lack of knowledge but of implementation. With its technological advances in renewable energy, Germany has no reason to fear changes as a result of 100-percent solar energy transition. On the contrary, today's resources crisis offers us a unique chance: alternative energy technology.

For the first time, the breakthrough can ensure that our children and grandchildren look to the future with self-confidence, optimism and joy. One day we should be able to say: 'Children, this is your world; we have helped to make life beautiful, for you as well. Life is waiting for you.'

We adults must now learn from children. We might finally grow up. It is high time everyone who thinks he is a grown-up oriented him or herself by the realism of the Fridays for Future movement. This means that each individual assumes responsibility, goes on strike, becomes politically active, learns to take him or herself seriously and stops being childish; in other words, adults must finally become grown-ups. No more illusions of eternal growth. Growth for growth's sake is an idle objective. An ecosocial market economy strives for sustainability and quality of life for everybody. Consider questions such as this: What is more important, healthy and sufficient nutrition for all or even more cars, mobile phones and intercontinental air travel? It is time to face realities.

2. Is there still hope for us?

The Dalai Lama suggests that climate policy follows science, just as the Fridays for Future movement does. Young people have nothing to lose but their future. Tell me, what is more important than the future of our children and grandchildren?

Is there still hope for us? Yes, there is. But not for long!

Our transformation is not blind fate. The future is what we make of it today. The most effective way to predict the future is to shape it. Looking back on my life, I would say that the ecosocial market economy is the most effective system in which billions of people can realise their dreams of a better world.

The Paris Climate Agreement of 2015 as well as the UN Millennium Development Goals (MDGs) are the basis of this new global ecosocial market economy. As a realist and journalist, I know of course that there is a big difference between writing about wonderful targets and meeting them. Real evolution means change, modification, transformation, turnaround and future. There is much to do. Future means working for the future. And all this must be beautiful, aesthetic, attractive – and not a deterrent.

The last time a general strike took place in Germany was on 12 November 1948. 'It's in your life's interest. Join in.' was the motto on the posters of the trade unions. That general strike triggered the social-market economy in Germany, which from 1950 and 1980 promised 'prosperity for all'. The promise was kept in a remarkable way. This historic strike became the basis of the German economic miracle. Shortly afterwards, Ludwig Erhard and his administration re-enforced social insurance schemes and introduced price regulations.

The global climate strike on 20 September 2019 could be the starting signal for a worldwide social-ecological market economy. Perhaps the upheaval caused by the global pandemic could even hasten its arrival. In 163 countries, more than four million people took to the streets. They not only wrote history. Perhaps even more important, their posters bore the words, 'We will come back.'

This first worldwide strike in human history was about nothing less than saving the world. On the same day the German government decided to apply climate protection measures – albeit at a minimum level – and three days later the UN met to discuss climate protection. Never before had there been such a weekend: the world is rising up. The issue was finally where it belonged, at the centre of international politics. Millions of young people took to the streets – and they will do again. The rulers of the world are under pressure: in Australia and India, in Germany and France, in the USA and in Bolivia, in Kenya, Bangladesh and in South Africa.

After World War II, the fathers and mothers of the social market economy in Germany were courageous and forward-looking. The well-known phrase of the French Enlightenment thinker René Descartes, 'I think, therefore I am' should today be altered to take into account the Buddhist enlightened insight – 'I live compassionately, therefore I am' – or the message of Carl Gustav Jung: 'I dream, therefore I am'. Most

Indians, Chinese, Africans or South Americans will never understand 'I think, therefore I am' – such an important phrase to we Europeans – any more than they understand our European atheism. Most people do not define themselves through reason but through emotion, not through 'me' but through their relationship to fellow beings, or through 'us'.

Buddha and Jesus were the most important, most convincing and most lasting role models of the past two and a half millennia. However, we have not learned enough from them. Otherwise, we would not upgrade nuclear weapons, wage wars or destroy the environment. Who ultimately prevents us from learning from their teachings if not ourselves?

What we can learn in Buddha's ecological teaching, as well as in Jesus's ecological teaching, is trust in the Creation. Jesus's decisive question for each of us: Do you trust in money or in God? In greed or love? His basic programme is the Sermon on the Mount. And Buddha's basic programme is compassion for all life, which he teaches us in the Eightfold Path.

Today's prevailing, near-global and unbridled materialistic neoliberalism has in reality become a dictatorship of international financial capital. The

Dalai Lama comments that money is an important means of exchange, but 'it is wrong to consider money a god or a substance endowed with some power of its own'. Meanwhile, Peter Spiegel wrote the book, *WeQ—More than IQ. A Farewell to the I-Culture*, describing a new 'we-economy'. The 'we-economy' shows new ways towards a more humane economy based on greater mindfulness, empathy, nonviolence, truthfulness, transparency and responsibility. These six virtues are the basis of the ethics of compassion, as advocated by Buddha and the Dalai Lama. These Buddhist economic ethics are similar to what Hans Küng, a Catholic theologian, calls the 'World Ethos' or the Protestant theologian Albert Schweitzer, 'Reverence for Life'.

There are already an increasing number of groundbreaking WeQ trends and projects, which I am going to outline in cooperation with Peter Spiegel in a separate book. We will point out that this revolution of compassion is nothing but active altruism. That is to say: the gain of others pleases me as much as my personal gain. This opens our hearts. Such altruism is spiritually motivated. Supporting the well-being of all living creatures naturally includes my personal well-being. An economic theory that ingenuously identifies happiness with material wealth is at best naive.

As a mendicant monk, the Dalai Lama has practically no personal belongings. Nevertheless, he is regarded by millions as their happiest contemporary. Every day this religious leader meditates for up to four hours. Training the mind. That is: peace of mind, happiness of recognition and understanding of our illusions. 'Inner tranquillity is actually the source of happiness,' according to economics professor Karl-Heinz Brodbeck.

20 September 2019 can become a turning point in human history, or at least the beginning of a turnaround. Climate protection, climate justice and solidarity take on a new meaning.

Is it all an illusion? Who, in 2018, would have dared to predict that a Swedish teenager was about to bring new inspiration to the agenda of world politics?

Acknowledgements

In writing this book, the Office of His Holiness the Dalai Lama in Dharamshala, India, and the Gaden Phodrang Foundation of the Dalai Lama in Switzerland have been of invaluable help. I am very grateful that, for more than 35 years, I had the chance to inform the audience about environmental issues on my radio and TV station, SWR, but also on Arte, 3sat, and the Third Programmes. This information often had to be pushed through against the will of the ARD authorities. Today, more colleagues than ever before keep raising these survival issues, for instance, Volker Angres, Harald Lesch and Sven Plöger.

In 1981, Bigi Alt secretly shot a TV movie in Tibet without Chinese watchdogs. This film was broadcast by Franz Alt several times on German (ARD) and foreign TV stations. And this was the beginning of a lifelong friendship between the Dalai Lama and the Alt family.

For more information: *www.dalailama.com* and *www.sonnenseite.com*

Index